W9-BPJ-751

—The Soul of Nature—

Celebrating the
Spirit of the Earth

EDITED BY
Michael Tobias
and
Georgianne Cowan

Ⓟ
A PLUME BOOK

PLUME
Published by the Penguin Group
Penguin Books USA Inc., 375 Hudson Street, New York, New York 10014, U.S.A.
Penguin Books Ltd, 27 Wrights Lane, London W8 5TZ, England
Penguin Books Australia Ltd, Ringwood, Victoria, Australia
Penguin Books Canada Ltd, 10 Alcorn Avenue, Toronto, Ontario, Canada M4V 3B2
Penguin Books (N.Z.) Ltd, 182–190 Wairau Road, Auckland 10, New Zealand

Penguin Books Ltd, Registered Offices: Harmondsworth, Middlesex, England

Published by Plume, an imprint of Dutton Signet,
a division of Penguin Books USA Inc.
This is an authorized reprint of a hardcover edition published by Continuum.
For information address the Continuum Publishing Company,
370 Lexington Avenue, New York, New York 10017.

First Plume Printing, April, 1996
10 9 8 7 6 5 4 3 2 1

Copyright © Michael Tobias and Georgianne Cowan, 1994
All rights reserved

 REGISTERED TRADEMARK—MARCA REGISTRADA

LIBRARY OF CONGRESS CATALOGING-IN-PUBLICATION DATA
The soul of nature : celebrating the spirit of the Earth / edited by Michael
    Tobias and Georgianne Cowan.
        p.   cm.
    Originally published: New York : Continuum, c1994.
    ISBN 0-452-27573-3
    1. Environmental responsibility.   2. Environmental ethics.
I. Tobias, Michael.   II. Cowan, Georgianne.
GE195.7.S68   1996
304.2—dc20   95–46474
                                                          CIP

Printed in the United States of America

Without limiting the rights under copyright reserved above, no part of this
publication may be reproduced, stored in or introduced into a retrieval system, or
transmitted, in any form, or by any means (electronic, mechanical, photocopying,
recording, or otherwise), without the prior written permission of both the copyright
owner and the above publisher of this book.

BOOKS ARE AVAILABLE AT QUANTITY DISCOUNTS WHEN USED TO PROMOTE PRODUCTS
OR SERVICES. FOR INFORMATION PLEASE WRITE TO PREMIUM MARKETING DIVISION,
PENGUIN BOOKS USA INC., 375 HUDSON STREET, NEW YORK, NEW YORK 10014.

(The Acknowledgments on page 10 constitute an extension of this copyright page.)

CELEBRATE OUR MYSTICAL CONNECTION TO THE EARTH IN *THE SOUL OF NATURE*

Gathered in this diverse anthology are some of the most cherished and eloquent voices from the eco-spirituality movement. Here are more than thirty essays—both passionate and reflective—from pivotal writers and thinkers of our time:

"I would like to learn, or remember, how to live. I come to Hollins Pond not so much to learn how to live as, frankly, to forget about it. That is, I don't think I can learn from a wild animal how to live in particular . . . but I might learn something of mindlessness, something of the purity of living in the physical senses and the dignity of living without bias or motive."   —from "Teaching a Stone to Talk" by Annie Dillard

"Now I step away from the ruins, toward the cliff's edge, and seat myself at the center of a sunken bowl of earth circled by a tumbled wall of sandstone blocks. . . . I brush out my tracks in the dust of the alcove floor—a personal ritual of respect—hug back along the cliff-hanging trail, ease over the crumbled defensive wall, scramble back up the talus slope, pull myself up through the split to the canyon rim, and take the scenic route home—out here in this magical landscape of rose and buff sandstone, azure sky and blue-green piñon and juniper, every route is the scenic route home." —from "A Place Where Spirits Dwell" by David Peterson

"Ecology is the mysterious work of providing a home for the soul, one that is felt in the very depth of the heart. Once we have the imagination that sees home in such a profound and far-reaching sense, protection of the environment will follow, for ecology is a state of mind, an attitude, and a posture that begins at the very place you find yourself this minute and extends to places you will never see in your lifetime."
          —from "Ecology: Sacred Homemaking" by Thomas Moore

MICHAEL TOBIAS, Ph.D., is the author of more than twenty previous books, including *Voice of the Planet*, which was made into a 10-part miniseries on TBS. He has written, directed, and produced more than seventy films for PBS, MacNeil-Lehrer, and other organizations.
GEORGIANNE COWAN is the director of the Spirit and Nature program at the Earth Trust Foundation, a nonprofit environmental education organization in Malibu, California. She has given workshops for more than fifteen years on creativity and nature, and is the producer of the video *Earth Dreaming*.

With love
and appreciation, to
Jane Gray Morrison
and Charles Bernstein

—Contents—

Part III: Living Every Day

Part IV: The Future of Nature

—Acknowledgments—

"Alighting upon the Daurian Steppe: A Mongolian Journey in Search of the White-Naped Crane" by Peter Matthiessen, reprinted from *Harper's Magazine*, June 1993, pp. 47–55.

"Love in Action" by Thich Nhat Hanh, reprinted from *Love in Action: Writings on Nonviolent Social Change* by Thich Nhat Hanh (1993) with permission of Parallax Press, Berkeley California.

"People, Land, and Community" by Wendell Berry, reprinted from *Standing by Words* by Wendell Berry, Copyright 1983 by Wendell Berry. Reprinted by permission of North Point Press, a division of Farrar, Straus & Giroux.

"Creation Spirituality" by Matthew Fox, reprinted from *Resurgence Magazine*, Ford House, Hartland, Devon EX39 6EE, UK.

"Creating an Ecological Economy" by Petra Kelly, reprinted from *Thinking Green: Essays on Environmentalism, Feminism, and Nonviolence* by Petra K. Kelly (1994), with permission of Parallax Press, Berkeley, California.

"The Age of Light: Beyond the Information Age" by Hazel Henderson, reprinted from *Paradigms in Progress: Life Beyond Economics* by Hazel Henderson, chap. 9, pp. 261–73. Published by Knowledge Systems, Inc., Indianapolis, 1991.

"Pilgrim at Tinker Creek" by Annie Dillard. Copyright 1974. *Flood* excerpt, *The Horns of the Altar* excerpt. Reprinted by permission of HarperCollins Publishers.

"Teaching a Stone to Talk" by Annie Dillard. Copyright 1982. *Living Like Weasels* excerpt. Reprinted by permission of HarperCollins Publishers.

"Holy the Firm" by Annie Dillard. Copyright 1977. *Newborn and Salted* excerpt. Reprinted by permission of HarperCollins Publishers.

"Apologia" by Barry Lopez. Copyright 1990 by Barry Lopez. Reprinted by permission of Sterling Lord Literistic, Inc.

"Spirit in Action" by Angeles Arrien. Excerpted from the speech given at the *1993 NAPRA Authors Breakfast* in Miami and from Angeles Arrien's recent book *The Four Fold Way; Walking the Paths of the Warrior, Teacher, Healer, and Visionary* from HarperSanFransisco.

"Island" from *Islands, The Universe, Home* by Gretel Ehrlich. Copyright (c) 1991 by Gretel Ehrlich. Used by permission of Viking Penguin, a division of Penguin Books USA Inc.

"Undressing the Bear" from *An Unspoken Hunger: Stories From the Field* by Terry Tempest Williams. Copyright (c) 1994 by Terry Tempest Williams. Reprinted by permission of Pantheon Boks, a division of Random House, Inc.

"On Sustainability" by B. D. Sharma. Reprinted by permission of *Sanctuary Magazine* and its editor, Bittu Sehgal, Bombay, India.

—Preface—

In myriad ways, we human beings have been compelled to articulate our feelings about "nature." The importance of this impulse has gained unprecedented status in the present century, such that our concerns for the environment are coming to dominate and shape most national priorities, cultural mindscapes, dreams, and fears. The profusion and diversity of viewpoints concerning nature are what is truly worth talking about. Collectively, our ecological attitudes, psychological deep structures, artistic aspirations, and spiritually-motivated understanding, constitute our best hope for reversing the devastating impact of our kind on the existing world. The quest for natural mind, natural homecoming, the wish to be re-attuned to the imperatives of nature, is a matter of ultimate and urgent importance. Our countless "solutions"—be they technical, social, economic, political, and so on—will lead us out of the melancholic labyrinths of ecological degradation only if human nature can muster the strength and wisdom to shed its accumulated bad habits. Some see this as an impossible hat trick, much like altering evolution. Others find the whole odyssey toward wildness to be ineffable, a Wittgensteinian exercise in silence, a personal and symbolic system better left to the inner world of the spirit. Philosophical pessimists might contend that humanity's nostalgia for all that is deemed "natural" is mere solipsism, unresolvable, forever divorced from real nature. Yet, there is an equally plausible case to be made for human reflection and expression as nature herself, self-yearning, self-realizing. According to this view, the ideal of nature seeks to perpetuate itself in every living organism. Meditate on nature and you invite the soul to partake of and become a universal truth.

All of these mysterious, sometimes plain-speaking, other times obscure, dimensions of the human connection to the wild are observed and summoned in this volume. While some of the essays have been published elsewhere, most are original. The editors have invited contributors with an eye toward suggesting a vital diversity of viewpoints, all focused on ecological themes. The many essays presented here spring from numerous disciplines: economics, aesthetics, biology, natural history, politics, poetry, mythology, feminism, anthropology, indigenous spirituality, environmental activism, comparative religion, human and non-human animal rights, prayer and meditation, storytelling, and psychology. What unifies this interdisciplinary feast of thought is a central focus, namely, the human desire to know ourselves as "nature" in its essence, and by so doing, to heal a troubled world.

The Soul of Nature offers no conclusion. It is, by definition, a work in progress, compassed by evolving lives, by poetic expression that cannot be measured against any time-line. The search for meaning, rather than any bold certitude, permeates these many patches of a colorful quilt. As members of a so-called "species," the authors presented here are asking questions, stopping to admire natural phenomena, or pointing out ill-fitting human behavior. The resulting tapestry offers hope and pellucidity, amid an often unsettling profile of humanity at the end of the millennium.

Whatever the soul of nature is, however it be felt and viewed and expressed, there is a little doubt that twentieth-century Homo sapiens are more compellingly tied to the possibilities and responsibilities of that soul than ever before. Environmental restitution, like all cleansing and remediation, invokes a practical activity. But no resolution can be sustained, or sustainable, without a corresponding inner process of clarification, soul-searching, and commitment. These endeavors are inextricable.

Ultimately, what the human being knows, nature knows. What nature wants, the human being must also want. Together, organisms rooted to the same tree of life, we will plummet, muddle, or ascend, depending upon the strength of our convictions and belief in the natural process. Humanity now finds itself stemming a chasm, an ecological abyss. Nature is blind before our avalanche. Our sheer numeric increase, insistent development, and short-term motivations have all but sapped the resurgent possibilities of millions of other species and countless habitats. Strength of human character and virtue, and the willingness to collaborate on selfless strategies of peace will spare the world, if we are resolved to do so. This wish for a gentle, biodiverse future, underscored throughout the present book, is itself an expression of the soul of nature in the very process of making itself known to human cognition.

—*Michael Tobias*

I t is midsummer. Outside, the full moon is rising. The cycle of one year has passed and in that time *The Soul of Nature* has matured into a ripe fruit with a bevy of chambers. Inside the body of this work, uncommon strains of seed are harbored; seeds of inspiration, of hope, of alarm, and of reckoning.

In re-reading the contributions contained within this potent volume, one is struck by the passion and commitment that each contributor brings to his or her topic. Revealed in their writing is a very *particular* and *private* relationship that each author has developed with nature. Often, a spirit of place emerges, forming the tableau from which their stories have been spun. There is humility and reverence contained within these pages—untarnished truths.

The contributors, as our guides into a sacred and mysterious realm, invoke a place where the personal becomes universal. Their words invite and hasten us to return to, and celebrate our home. They speak of ancient bloodlines that run deep, charging through our veins, and reminding us that we *are* related, to each other and to this primal legacy called nature, called earth.

—Georgianne Cowan

The inception of this book arose out of Earth Trust Foundation's *Spirit and Nature Speakers Series*, directed by Georgianne Cowan. Many of the contributors to *The Soul of Nature* have presented their ideas within the context of this series. Others were chosen by the editors because of the integrity of their writing addressing themes concerned with spirit and nature. Michael Tobias, the first speaker in this series, germinated the idea of assembling the anthology. *Anthology* literally means flower-gathering—a fitting definition for a bouquet of writings from some of the most inquiring and poignant authors on the subject of nature.

Earth Trust Foundation is a non-profit organization which encourages personal awareness and social responsibility from a global perspective. Started in 1985 by Andrew Beath, Earth Trust Foundation is a catalyst for community involvement in social action to benefit the environment and the human condition. Through its local and international Earth Ways programs such as *The Rainforest Preservation Campaign*, *John Seed Directed Grants*, *Women's Village Banking*, and the *Spirit and Nature Speakers Series*, Earth Trust Foundation has developed a community of socially and environmentally concerned people working to facilitate ecological preservation. Net proceeds from the sale of this book will go towards funding Earth Trust Foundation's environmental programs—*helping to preserve the sanctity of all life*.

For further information write: Earth Trust Foundation, 20110 Rockport Way, Malibu, CA 90265. Telephone: (310) 456-8300.

Part I

—Earth Sapiens—

—1—
A Place Where Spirits Dwell

David Petersen

It's night down in the sandstone canyons of southeastern Utah, though twilight still lingers up here on their rims, where my little camp is set near a vertical stone gash plunging a thousand feet into . . . what? I hope never to find out. I've come here to this ancient stone heart of the Southwest alone, pack on back, as I do every once in a while, seeking recreation, relaxation, solitude. Nothing more, nothing less.

As I sit gazing into a sparking campfire, it occurs to me that to some eyes, the rocky, semi-desert landscape that comprises much of the Colorado Plateau must seem desolate, foreboding, more dead than alive. Not true; it's just that life hereabouts doesn't flaunt itself. Visible to anyone who cares to look are a wealth of diurnal creatures: the turquoise sky virtually swarms with bird life: darting cliff swallows, bell-voiced canyon wrens, swifts graceful in aerobatic flight, hawks, falcons, eagles, vultures, blue-black iridescent ravens; while on the ground, the shade of pinon, juniper, and sage hides deer, rabbits, ground squirrels, lizards, snakes, the stodgy desert tortoise, and myriad others.

Come night, you need only listen to discover life. Even now, as I sit here fireside and muse, a great-horned owl's persistent query—*Who, who are you?*—animates the night. The owl's call is answered by a melancholy nightjar—*Poor-will, Poor-will*. Some small, unseen mammal scurries in the sagebrush beyond the campfire's little cave of light. And always, above, around, and through all other sounds, rings a cacophony of crickets.

I let the fire burn to coals, roll out my nylon sleep-sack on the warm soft sand and prepare for the "little death." No tent, and no need for one; out here, on a sublime spring night such as this, the luminous heavens are shelter enough. I fall asleep to the incense fragrance of smoldering pinon pine.

Waking to a morning made for hiking, I strike off south along an undulating slickrock canyon rim of eroded sandstone, a terrain looking for all the world like an ocean of petrified waves. My destination, a couple of miles distant, lies hidden beneath a promontory overlooking a broad canyon floor a quarter-mile below. To find the place, you have to know it's there. A long time ago, a friend led me here. Keeping the faith, I've shown it to only one other—a secret too widely shared loses its magic.

At the promontory, I lower myself through a split in the rim rock, then skid and slide down a steep talus slope to a narrow ledge a hundred feet below. Hugging the ledge as I cross the vertical cliff face, trying not to look down, I emerge, weak and shaky, at the mouth of a hidden alcove defined by an overhanging cliff above, a sheer vertical drop below and a sloping shelf eroded a few yards back into the cliff face.

This is a place of spirits.

Three hundred years before Columbus sailed, an extended family, or clan, of Stone Age Indians lived in this west-facing rock shelter. The Navajo who occupy this arid land today call them *Anasazi*—"the ancient ones." In addition to building sturdy pueblos of adobe and hand-shaped sandstone blocks, the Anasazi hunted, gathered wild foods, farmed, and fired fine painted pottery, black on white. As mute evidence of their skill and prolificity as potters, palm-sized shards of mugs, jugs, bowls, and ladles lie strewn about the alcove floor. I pick up a few pieces, admiring their geometric designs, yet clean and sharp, refuting seven centuries of wind, rain, and sun. One at a time, I return the shards to their rightful place in the enchanted dust.

Amazingly, a few desiccated corncobs also survive, dehydrated and shrunken by the arid climate. These are the produce of a crop grown and harvested even as Kublai Khan and his Mongol raiders ran amok through Asia . . . even as a dreary Europe suffered through the horrors of its Middle Ages.

What, I wonder, might have prompted these mysterious people to nest on such risky aeries as this and hundreds of similar others all across the Four Corners region of the Southwest? The daily chores of hauling food, water, and firewood along that damnable ledge would have been onerous, even life-threatening. You could never allow your children out of sight, or hand, for even a moment.

So, why *did* this little clan of Pueblo Indians live here? The strongest surviving clue is a half-ruined defensive wall erected across the narrowest span of the approach ledge. A portal in its center is just large enough to allow one person at a time to squeeze through. A lone sentry stationed out there with a club could easily have beaten back an army of would-be raiders as they attempted, one fool at a time, to squeeze through the portal.

Obviously, like the overwhelming majority of Anasazi cliff dwellings elsewhere, this place was selected for its defensive capabilities, designed and built as a fortress against unknown ancient enemies . . . raiding Utes down from the flanks of Colorado's silver San Juans? Internecine rivals from neighboring Anasazi clans? We shall never know for certain. Nor, so far as I'm concerned, should we.

Perhaps, as well, perching their homes on such scenic aeries held some spiritual meaning for the Anasazi. We shall never know that for certain, either. Nor, so far as I'm concerned, should we.

I stoop into one of the four almost identical low-roofed rooms—rectangular, just large enough to accommodate a couple of adults and maybe a child or two. I find that I can sit comfortably, but standing is impossible; the ceiling—which consists of nothing more than the sloping alcove roof—is quite low. A single small door/window looks west across the canyon. Fire-blackened inner walls and ceiling confirm that this tiny chamber was used as a living space. A row of wooden pegs spaced at about six-inch intervals jut from the crumbling mud-plaster high along the inner front and side walls—pegs from which once may have hung buckskin and woven-fiber bags containing valued personal belongings, talismans, ceremonial magic. Or perhaps the pegs supported a narrow shelf. Probably both.

Feeling dusty and claustrophobic inside, I crawl back out into the brilliant southwestern sunshine and continue my reverent poking around.

Tucked back into the narrow, pinched-off rear of the alcove at either end of the row of connected apartments are two rounded, bat-guano-glazed granaries. One of the pair has an inch-thick, hand-shaped sandstone slab—a door—leaned against its small opening. In the dust of the granary floor lie scattered a few mummified corncobs and the black hard pellets of wood rats.

Now I step away from the ruins, toward the cliff's edge, and seat myself at the center of a sunken bowl of earth circled by a tumbled wall of sandstone blocks. Beneath me (I know from having seen many such places after excavation) lies a collapsed kiva, an underground ceremonial chamber. In this little pueblo's heyday, down in this dark, smoky cellar, the men of the clan gathered nightly to talk, sing, smoke, and perform magical ceremonies. This we *do* know for certain, for even today the Pueblo peoples of New Mexico and Arizona—direct descendants of the Anasazi—practice similar, perhaps identical rites down in similar, perhaps identical kivas.

Seduced by the haunting atmosphere of this place, I consider spending the night here, but quickly think better of it. There is no firewood, I have no food and not enough water. Besides, respect demands leaving these musty old ruins to the juniper-scented ghosts of the people who, for reasons as recondite as those that brought them here in the first place, one day just walked away from it all, deserting a native homeland considered by them as *Sipapu*—the womb of humanity.

For whatever reasons, the Anasazi are gone. Forever. And I too must leave.

I brush out my tracks in the dust of the alcove floor—a personal ritual of respect—hug back along the cliff-hanging trail, ease over the crumbled defensive wall, scramble back up the talus slope, pull myself up through the split to the canyon rim, and take the scenic route home—out here in this magical landscape of rose and buff sandstone, azure sky and blue-green pinon and juniper, *every* route is the scenic route home.

I sag into camp just at dusk, drop my pack, and slump down to the living heart of this arid place—a drip-spring hidden in a shaded grotto at the head of a small side canyon. Uncapping my two parched canteens, I place them on the sand beneath a slow line of silver droplets emerging from a seam in the sandstone wall. There is no sweeter music, as my old friend Edward Abbey was wont to say, than the *tink-tink, tink* of desert water dripping into a tin cup. And there is no sweeter taste, I am wont to say, than cool spring water and a splash of George Dickel sipped within the perfumed smoke of a campfire in the American Southwest on a mild spring night.

As the day dims and the lazy spring drips, drips, in no hurry whatsoever to satisfy me, I stand gawking in wonder at this place. Here, as in uncounted similar oases flung by geologic happenstance across the Colorado Plateau, grow anomalous riparian plant communities utterly dependent for their survival on scant moisture leaking improbably from "solid" rock. Like the biblical burning bush, desert drip springs are miracles in the wilderness—miracles you can drink. Carpeting the damp grotto wall along and below the drip line are lush mosses. And thriving in and around the pellucid spring pool below are cattails, bracken ferns, Indian rice grass, and once saucy, red-lipped monkey-flower.

At my feet, the damp sand rimming the pool is a journal of recent wildlife activity. From the tracks I read that a rabbit, various smaller rodents, a fox, and an adult mule deer have been here recently. The usual lot of thirsty desert mammals. And that's not all. In the periphery of my vision I notice an odd depression and step over to investigate. My pocket flashlight cuts through the dimming twilight, and I see a track as big as a man's hand—much larger than any coyote, though not so big as a bear—and rounded, with no claw marks visible. The bi-lobed front edge and tri-lobed rear of the wide plantar pad are clearly evident . . . cougar. The single print is sharp-edged, obviously fresh. I look around but can find no others; apparently, the cat ventured just his one step off the slickrock toward the spring, then withdrew.

My skin prickles with the realization that one of the largest and most sublime predators in North America has been here, right here, and not all that long ago . . . a beast of the magical clan my pal Doug Peacock—no stranger to canyon country—refers to as "charismatic megafauna."

I drop to my knees and study the track from every angle, then stand again and point the little flashlight all around. But the batteries are weak and the tired yellow beam doesn't reach far. In another minute or so, both twilight and flashlight will fade completely, and I'll be left here in the dark. Alone . . . or perhaps not.

Be cool, man. Like the old saw says, there's nothing to fear but . . . what? Why is it we tend to fear the unseen more than the visible, the unknown more than the known?

Statistically, I've heard, we are several hundred times less likely to be attacked by a mountain lion than to be struck by lightning. May be. But there isn't a storm cloud in sight and the biggest cougar track I've ever seen is fresh at my feet and my heart is pounding in my throat.

I snatch up the two sloshing, half-filled canteens in one hand—the quart they hold between them will have to suffice until morning—clutch the impotent flashlight in the other hand and feel my way back up to camp and the comforts of home and hearth.

After replacing the spent flashlight batteries with spares from my pack, I busy myself with evening camp chores: spread ground-cloth, sleeping pad and bag on the sand near the fire pit, kindle a blaze, scorch and devour a couple of elk steaks brought from home—frozen yesterday but beyond thawed now.

Much later, I toss one last club of pinon onto the coals, then strip and slide into the cocoon security of my sleep sack. Just me, ten billion desert stars above and two little brown bats turning and diving after the gray moths circling at the twinkling edges of firelight. Thinking still of that big round track down by the spring, sleep is a long time coming.

But sleep I do, until—along about midnight—a family of coyotes wakes me, yammering maniacally, sounding close but probably not close at all. The fire is dead and all the world is as black as magic. I manage to keep my eyes open just long enough to witness a blue-green comet make a futile rush for the western horizon.

When the dream comes, it is almost palpable—no clear images, only eerie, suspiring, susurrus sounds like the guarded footfalls of a prowler . . . like rapid, rhythmic breaths. Feeling vaguely threatened by the ethereal sounds, I awaken and sit up. Though I'm almost too warm in my bag, I notice that my arms are pimpled with gooseflesh. I peer into the darkness but see nothing. I listen but no sound comes. All is quiet in the anthracite desert night. The stars have dimmed and even the hooters and crickets have fallen quiet. I consider switching on the flashlight for a look around but don't, for fear I'll think myself a coward come morning. Finally, still uneasy and feeling a little foolish, I lie back and hope for sleep to return. Or for morning to come.

In the soft glow of dawn I wake all bleary-eyed and groggy to discover that my eerie dream was in fact no dream at all. There in the powdery sand, just a body's length out from my sleeping place and imprinted over one of my own boot marks, is a big round track.

I unzip my bag and stand, naked, and peer around. Prints are everywhere. Over there the lion approached from the sage. And those odd marks show where he, or she, sat back on lean haunches, long tail sweeping a fan-shaped arc in the sand. From that thoughtful repose I imagine the prowler staring at me with big nocturnal eyes, listening, panting with fast shallow breaths, pondering the redolent sleeper in its own mysterious feline fashion.

Near as I can read the spoor, it appears the lion then moved to the far side of the burned-out campfire and haunch-sat again. And over there, finally, its curiosity apparently satisfied—or maybe I startled it with my sudden awakening—the cat padded back into the sage. Perhaps it remains nearby, biding its time, biding mine.

Following a sudden and reckless urge, I pull on shorts, lace boots over sockless feet, and follow the departing trail. But the prints soon strike slick-rock and that's that.

Warmed by strong camp coffee and relaxed under a brilliant morning sun, I sit and reflect. Had the cat been looking to make a meal of me, it probably could have. Pumas are predation perfected, capable of bringing down not just deer, but creatures as large as elk and cattle and horses . . . and on very rare occasion, people. Pouncing after a short rush from hiding, the cougar kills by sinking long canines into the skull or neck of the surprised victim, then holding, holding. A naked sleeping man would be cake.

Perhaps predation was never the prowler's intent. Or, drawing near, the keen-nosed animal was offended by my second-day trail smell. Probably the former. Most likely, I spooked it from the spring last night and it's been lurking nearby ever since, curious as a cat. Was I in any real danger? I'll never know for certain. And just as well. For with the knowing would come the death of mystery, the end of magic.

I am not a spiritual person in any orthodox sense. One life at a time, please. Yet, out here in the ancient dust, among nocturnal dream creatures— out here in this place where spirits dwell—I feel that my life has been touched by magic.

—2—
Nowhere Ridge

John Murray

It is not down on any map; true places never are.

—Herman Melville

I lay down in the rock bunker and found it served a useful purpose. Whoever carried the talus to the knoll understood the effects of wind at twelve thousand feet. The structure faced west, along the axis of the prevailing Pacific storm track, and was shaped like the entrance to the womb, with the enclosed end pointing toward the wind and the open end toward the rising sun. Laying inside the shelter with my hands behind my head and my feet sticking out the entrance, I looked up and saw only the cold Colorado sky. Passing clouds. A curious gold eagle. The chirp of a pika. The sweet fragrance of high altitude clover. The blue curve of infinity.

What was this thing? Certainly not a frost heave or meandering rock channel. Too symmetrical for that. Nor was it a cairn. Too elaborate to mark a trail, and I was miles from the nearest trail. How old was it? The gray hoary chunks of granite, at least those at the base, had rested *in situ* for centuries. The lichens, which grow more slowly than human consciousness, attested to that. Dull iridescent red, they covered the foundation cobble like dried blood. Who built it? Yes, that was the question. Whoever or whatever, I could have remained awhile. For one thing, the grassy bed was well-suited for a nap. For another, the thin mountain air always acts like a narcotic, especially after ascending two thousand feet in about an hour. But I had another mountain, further off, to climb that bright August day. I stood up, walked around, scratched my head in befuddlement, dug the camera out of the pack, and took a picture. The reconnaissance complete, I methodically surveyed the horizon, as I had been taught in the Marines long ago, to orient myself should it become necessary to make a report.

To the west were the Never Summer Mountains, the same classic range that Ansel Adams memorialized in a well-known photograph during the Great Depression. Mount Stratus (12,520'). Mount Nimbus (12,706'). Mount Cumulus (12,725'). Mount Cirrus (12,797'). The supreme work of some thirty million years was subsequently mutilated by the Denver Water

Board and the Grand Water Ditch, a monumental act of hubris that
diverted the headwaters of the Colorado River to the parched East Slope.
Imagine Thomas Moran's 1875 painting *The Mountain of the Holy Cross* with a
well-engineered gash running the width of the canvas. Aeschylus comes to
mind: "Those that the gods would destroy, they first make mad with
power." To the north was the Mummy Range, on the far side of which you
will find a lost country—the Comanche Peak Wilderness—as beautiful as
anything I have seen in six years of exploring the Alaskan interior. To the
south was Middle Park. Cold clear streams falling swiftly out of the high
country. Low rocky ridges of quaking aspen and Douglas fir. A sportsmen's
paradise, where I fished and hunted in my youth, before the ski resorts, the
dam project, the clear-cuts, the summer home sprawl, and the general
tourist blight turned the wild democratic valley into Little Europe.

Below Mount Baker (12,397') I could just make out Baker Gulch,
where in 1973 a large male black bear killed and partially ate a Wisconsin
man sleeping in his tent. Ray Lyons, a former World War II bomber pilot
and West Slope outfitter I later worked for as a hunting guide in 1977, was
summoned from his Grand Mesa cattle camp to track down and kill the
bear. He and his Plott hounds (a modern version of the Greek Furies)
accomplished that task in one day. More recently, green activists successfully
halted the logging of old growth forest in Baker Gulch. To the east, where
the far ponderosa foothills dropped like headlands, was the vanished buffalo
prairie of the Pawnee and the Arapahoe, colored a light blue like the sea.
Somewhere beyond that oceanic horizon were the distant corn fields and
buckeye woodlots of Ohio where I spent a childhood looking West. The rising smog of rush-hour Denver coffee-stained the southeastern portion of the
compass. To those familiar with the Colorado Front Range, it has now
become apparent that I was somewhere in Rocky Mountain National Park.
The place? Let us call it Nowhere Ridge.

I did not tarry long that day, did not tear apart the sod looking for
immaculate arrowheads or upend ancient dignified rocks to locate that well-
preserved axehead or grinding stone. My ignorance was total, blissful. Without so much as a backward glance, I hurried on my way, singing all the songs
from Bob Dylan's *Blood on the Tracks* I could rememeber, not realizing like so
many discoverers, that I had even made a discovery. I dutifully climbed my
appointed mountain, sat cross-legged at the summit, found a bighorn sheep
skeleton on the descent, befriended a brown-feathered ptarmigan while eating
an orange, spotted a herd of elk in a nameless valley, watched them graze and
play and splash in a tarn through binoculars, and then trotted the five miles
out to the road and drove back to Boulder.

Sometime in late December, when the snow was ass deep to an elephant and the mountain wagon was hopelessly high centered in the driveway and all my neighbors were skiing to the grocery store, I decided to sort my slides from the previous summer. When the forgotten image flashed on the screen I puzzled over it for a long time. Outside the house snowflakes fell, and the wind howled. Belatedly, I made a duplicate and sent it to Glen Kaye, Chief Park Naturalist, along with a letter. He probably knew about it, I wrote, but just in case here was a photo and map. I wondered if it might be part of an Indian game drive system. The bunker was located in such a way that game could be pushed past it by drivers working at timberline. A hunter could wait, concealed from view, with a bow and arrow, a spear, or an atlatl. I'd done the same thing several times while bowhunting for mule deer in the Mount Evans Wilderness Area with my old college friend Greg Fife.

Within the week, a letter arrived. No record of the bunker existed, Kaye said. A team of archaeologists from Colorado State University and the National Park Service would be dispatched next summer as soon as the snow melted. Please inform no one of the exact location as undisturbed sites are rare and "so much pre-history in the park has already been lost." In a general way, as I am telling you, I told my roommates Dave Student (real name) and Jeff Swedlund, both of whom tied flies for a living, about the site. We were excited and happy; it was refreshing to know that Rocky Mountain still kept her secrets with 2.2 million annual visitors (now over 3.0 million). As I recall, John Gierarch and Ed Engle came over later in the week for dinner and read Kaye's letter. Both have since gone on to become pretty well-known nature writers; John with books like *Trout Bum* and *The View from Rat Lake* and Ed with his essay collection *Seasonal*. At the time, like so many nature writers who are actively publishing today, we were still dabbling in a sort of transplanted Chinese nature poetry, unaware that there was this *other* genre growing faster than fireweed through a burned forest.

The results of the field study were sent to me about a year later, on the eve of their publication as a scientific monograph. As it turns out, the U-shaped structure was a relatively recent vision quest site built upon a prehistoric (10,000 years old) fasting bed. Both were used by Native Americans for meditative purposes. In some cases, as we know from Sitting Bull and Black Elk, self-mutilation with sharp obsidian blades was also practiced, the pain being seen as a way to further liberate the spirit. An idea of how far the participants would go is evident in George Catlin's 1837 painting *Self-Torture in Sioux Ceremony*, which depicts a Teton Sioux hanging from a pole by tongs connected to sharp spikes piercing the skin and muscle of his chest. This ritual was part of the Sun Dance, which sought to acquire power from

Wakan Tanka (The Great Mystery). Some of the last historic vision
questers were interviewed by George Bird Grinnel, the naturalist who
accompanied General Custer on his 1874 Black Hill expedition. Grinnel
wrote of the Cheyenne in his 1923 book:

> In those old times, young men used to go off on the hills and
> fast for four nights. This was call wú wŭn', starving. They did
> this in order that they might be fortunate, and might not be hit
> in battle . . . they had no shelter and no covering. . . . If he
> fasted to the end, after four days, the old man went to him and
> brought him down to the camp. They did this only in the sum-
> mer. This is said to have been purely a sacrifice, and not an
> attempt to dream for power; but often those who lay there did
> have dreams, and what they dreamed surely came to pass. Not
> everyone starved and to only a part of those who starved did the
> vision come.

As I had expected, there was also evidence that the site had been used in
connection with a game drive system. The archaeologists discovered sixteen
additional cairns on the ridgeline as well as a second, though much less evident, stone structure in the shape of a small circle. All totaled they
unearthed twenty-one arrowheads, seven butchering tools, 202 chipping
flakes, and two grinding slabs. The oldest of these was a white chert projec-
tile point dated to about ten thousand years before present, and the most
recent was a quartzite point that was about five hundred years old. The
artifacts came from quarries as close as Specimen Mountain in the Mummy
Range and as far away as lower Middle Park near present-day Kremmling
(where Louisiana-Pacific had until 1992 a facility that turned my favorite
aspen groves into toxic waferboard and their associated beaver ponds into
flooded quagmires). To read this monograph is to be humbled with the
knowledge that a vibrant human culture occupied the Colorado Rockies
when the ziggurats of Ur were still just a distant dream along the Tigris-
Euphrates.

Every summer since finding the site, almost ten years now, I have
climbed a mountain on the summer solstice, the longest day of the year, and
fasted from dawn to dusk. I take only water, a pencil and paper, perhaps a
book of poetry, and my foul weather gear. I think about the preceding year
and the year to come, what I have done right and what I have done wrong,
what is important and what is not, who remains and who has passed on.
Each year is different. Some years I come down exhausted. Other years I
return uplifted. I look not for the false euphoria of the cult follower or the

forced equanimity of the Stoic, both of which are illusions. I try only to better understand myself and my world. To see clearly, in the sense that the French philosopher Albert Camus understood the concept when he wrote in "The Wind at Djemila" that, above all, he wanted to "keep my lucidity to the last." To achieve balance and moderation as Aristotle used the terms. I believe the Plains Indians had a good idea in their pursuit of the vision quest as a mode of personal reflection. I also believe it would not be an entirely bad idea, in such a secular age, if more people did this. I certainly intend to take my son up to the hills and show him what I have learned when he reaches the age of awareness around thirteen or fourteen. The mountains have much to teach us, in their great resonant silences, in the spectacular beauty of their wide horizons, in the way the cloud shadows brighten and darken the ridges and valleys. To sit all day upon a heavy, solid mountain is to feel anchored. Do not get up and leave early. The more fatigued you become, the closer you are to achieving that which you seek. Stay until you take on the color of the earth and grow roots and sprout flowers that you can take back as gifts to share with loved ones, as evidence, to them and others, that there are worlds inside each of us that make the richest men and women seem idle fools and paupers.

—3—

Island

Gretel Ehrlich

I come to this island because I have to. Only geography can frame my mind, only water can make my body stop. I come, not for solitude—I've had enough of that in my life—but for the discipline an island imposes, the way it shapes the movement of thoughts.

Humpbacked, willow-fringed, the island is the size of a boat, roughly eighty-five feet by twenty, and lies on the eastern edge of a small man-made lake on our Wyoming ranch. I call this island Alcatraz because I once mistook a rare whooping crane that had alighted in the lower field for a pelican, and that's what the Spanish *alcatraz* means: pelican. But the name was also a joking reference to the prison island I threatened to send my saddle horse to if he was bad, though in fact *my* Alcatraz was his favorite spot on the ranch to graze.

Now Blue is dead, and I have the island to myself. Some days, Rusty, my thirteen-year-old working dog, accompanies me, sitting when I sit, taking in the view. But a view is something our minds make of a place, it is a physical frame around natural fact, a two-way transmission during which the land shapes our eyes and our eyes cut the land into "scapes."

I sit to sweep the mind. Leaves, which I think of as a tree's discontinuous skin, keep falling as if mocking my attempts to see past my own skin, past the rueful, cantankerous, despairing, laughing racket in my head.

At water's edge the tiny leaves of wild rose are burned a rusty magenta, and their fruit, still unpicked by birds, hangs like drops of blood. Sun on water is bright: a blind that keeps my mind from wandering. The ripples are grooves the needle of memory makes, then they are the lines between which music is written—quintets of bird song and wind. The dam bank is a long thigh holding all restlessness in.

To think of an island as a singular speck or a monument to human isolation is missing the point. Islands beget islands: a terrestrial island is surrounded by an island of water, which is surrounded by an island of air, all of which makes up our island universe. That's how the mind works too: one idea unspools into a million concentric thoughts. To sit on an island, then, is not a way of disconnecting ourselves but, rather, a way we can understand relatedness.

Today the island is covered with duck down. It is the time of year when mallards molt. The old, battered flight feathers from the previous spring are discarded, and during the two or three weeks it takes for the new ones to grow in, they can't fly. The males, having lost their iridescent plumage, perform military maneuvers on the waters, all dressed in the same drab uniform.

Another definition of the word *island* is "the small isolated space between the lines in a fingerprint," between the lines that mark each of us as being unique. An island, then, can stand for all that occurs between thoughts, feathers, fingerprints, and lives, although, like the space between tree branches and leaves, for example, it is part of how a thing is shaped. Without that space, trees, rooms, ducks, and imaginations would collapse.

Now it's January, and winter is a new moon that skates the sky, pushing mercury down into its tube. In the middle of the night the temperature drops to thirty-two below zero. Finally, the cold breaks, and soon the groundhog will cast a shadow, but not here. Solitude has become a reflex: when I look at the lake no reflection appears. Yet there are unseen presences. Looking up after drinking from a creek, I see who I'm not: far up on a rock ledge, a mountain lion, paws crossed, has been watching me.

Later in the month, snow on the lake melts off, and the dendritic cracks in ice reappear. The lake is a gray brain I pose questions to. Somewhere in my reading I come on a reference to the island of Reil. It is the name given to the central lobe of the cerebral hemisphere deep in the lateral tissue, the place where the division between left and right brain occurs, between what the neurobiologist Francisco Varela calls "the net and the tree."

To separate out thoughts into islands is the peculiar way we humans have of knowing something, of locating ourselves on the planet and in society. We string events into temporal arrangements like pearls or archipelagos. While waiting out winter, I listen to my mind switch from logic to intuition, from tree to net, the one unbalancing the other so no dictatorships can stay.

Now snow collapses into itself under bright sun with a sound like muffled laughter. My young friend Will, aged nineteen, who is suffering from brain cancer, believes in the laughing cure, the mango cure, the Molokai cure, the lobster cure—eating what pleases him when he can eat, traveling to island paradises when he can walk, astonished by the reversal of expectation that a life must last a certain number of years.

In the evening I watch six ravens make a playground of the sky. They fly in pairs, the ones on the left, for no reason, doing rolls like stunt pilots. Under them, the self-regulating planet moves and the landscape changes—fall to winter, winter to spring, suffering its own terminal diseases in such a way that I know nothing is unseasonal, no death is unnatural, nothing escapes a raven's acrobatic glee.

Excerpts

Holy the Firm, Pilgrim at Tinker Creek,
and *Teaching a Stone to Talk*

Annie Dillard

Holy the Firm

Every day is a god, each day is a god, and
holiness holds forth in time. I worship each
god, I praise each day splintered down,
splintered down and wrapped in time like a
husk, a husk of many colors spreading, at dawn
fast over the mountains split.

I wake in a god. I wake in arms holding my
quilt, holding me as best they can inside my
quilt.

Someone is kissing me—already. I wake, I
cry "Oh," I rise from the pillow. Why should
I open my eyes?

I open my eyes. The god lifts from the water.
His head fills the bay. He is Puget Sound, the
Pacific; his breast rises from pastures; his
fingers are firs; islands slide wet down his
shoulders. Islands slip blue from his shoulders
and glide over the water, the empty, lighted
water like a stage.

Today's god rises, his long eyes flecked in
clouds. He flings his arms, spreading colors; he
arches, cupping sky in his belly; he vaults,
vaulting and spread, holding all and spread
on me like skin.

Pilgrim at Tinker Creek

That something is everywhere and always amiss is part of the very stuff of creation. It is as though each clay form had baked into it, fired into it, a blue streak of nonbeing, a shaded emptiness like a bubble that not only shapes its very structure but that also causes it to list and ultimately explode. We could have planned things more mercifully, perhaps, but our plan would never get off the drawing board until we agreed to the very compromising terms that are the only ones that being offers.

The world has signed a pact with the devil; it had to. It is a covenant to which every thing, even every hydrogen atom, is bound. The terms are clear; if you want to live, you have to die; you cannot have mountains and creeks without space, and space is a beauty married to a blind man. The blind man is Freedom, or Time, and he does not go anywhere without his great dog Death. The world came into being with the signing of the contract. A scientist calls it the Second Law of Thermodynamics. A poet says, "The force that through the green fuse drives the flower/Drives my green age." This is what we know. The rest is gravy.

"In nature," wrote Huston Smith, "the emphasis is in what is rather than what ought to be." I learn this lesson in a new way every day. It must be, I think tonight, that in a certain sense only the newborn in this world are whole, that as adults we are expected to be, and necessarily, somewhat nibbled. It's par for the course. Physical wholeness is not something we have barring accident; it is itself accidental, an accident of infancy, like a baby's fontanel or the egg-tooth on a hatchling. Are the five-foot silver eels that migrate as adults across meadows by night actually scarred with the bill marks of herons, flayed by the sharp teeth of bass? I think of the beautiful sharks I saw from the shore, hefted and held aloft in a light-shot wave. Were those sharks sliced with scars, were there mites in their hides and worms in their hearts? Did the mockingbird that plunged from the rooftop, folding its wings, bear in its buoyant quills a host of sucking lice? Is our birthright and heritage to be, like Jacob's cattle on which the life of a nation was founded, "ring-streaked, speckled, and spotted" not with the spangling marks of grace like beauty rained down from eternity, but with the blotched assaults and quarryings of time? "We are all of us clocks," says Eddington. "whose faces tell the passing years." The young man proudly names his scars for his lover; the old man alone before a mirror erases his scars with his eyes and sees himself whole.

Through the window over my desk comes a drone, drone, drone, the weary winding of cicadas' horns. If I were blasted by a meteorite, I think, I could call it blind chance and die cursing. But we live creatures are eating each other, who have done us no harm. We're all in this Mason jar together, snapping at anything that moves. If the pneumococcus bacteria had flourished more vitally, if they had colonized my other lung successfully, living and being fruitful after their created kind, then I would have died my death, and my last ludicrous work would have been an Easter egg, and Easter egg painted with beaver and deer, an Easter egg that was actually in fact, even as I painted it and the creatures burgeoned in my lung, fertilized. It is ridiculous. What happened to manna? Why doesn't everything eat manna, into what rare air did the manna dissolve that we harry the free living things, each other?

An Eskimo shaman said, "Life's greatest danger lies in the fact that men's food consists entirely of souls." Did he say it to the harmless man who gave him tuberculosis, or to the one who gave him tar paper and sugar for wolfskin and seal? I wonder how many bites I have taken, parasite and predator, from family and friends; I wonder how long I will be permitted the luxury of this relative solitude. Out here on the rocks the people don't mean to grapple, to crush and starve and betray, but with all the good will in the world, we do, there's no other way. We want it; we take it out of each other's hides; we chew the bitter skins the rest of our lives.

But the sight of the leeched turtle and the frayed flighted things means something else. I think of the green insect shaking the web from its wings, and of the whale-scarred crab-eater seals. They demand a certain respect. The only way I can reasonably talk about all this is to address you directly and frankly as a fellow survivor. Here we so incontrovertibly are. *Sub specie aeternitatis* this may all look different, from inside the blackened gut beyond the narrow craw, but now, although we hear the buzz in our ears and the crashing of jaws at our heels, we can look around as those who are nibbled but unbroken, from the shimmering vantage of the living. Here may not be the cleanest, newest place, but that clean timeless place that vaults on either side of this one is noplace at all. "Your fathers did eat manna in the wilderness, and are dead." There are no more chilling invigorating words than these of Christ, "Your fathers did eat manna in the wilderness, and are dead."

Alaskan Eskimos believe in many souls. An individual soul has a series of afterlives, returning again and again to earth, but only rarely as a human. Since its appearances as a human being are rare, it is thought a great privilege to be here as we are, with human companions who also, in this reincarnation, are privileged and therefore greatly to be respected." To be here as we are. I love the little facts, the ten percents, the fact of the real and legged borers, the cuticle-covered, secretive grubs, the blister beetles, blood flukes and mites. But there are plenty of ways to pile the facts, and it is easy to overlook some things, "The fact is," said Van Gogh, "the fact is that we are painters in real life, and the important thing is to breathe as hard as ever we can breathe."

So I breathe. I breathe as the open window above my desk, and a moist fragrance assails me from the gnawed leaves of the growing mock orange. This air is intricate as the light that filters through forested mountain ridges and into my kitchen window; this sweet air is the breath of leafy lungs more rotten than mine; it has sifted through the serrations of many teeth. I have to love these tatters. And I must confess that the thought of this old yard breathing alone in the dark turns my mind to something else.

Teaching a Stone to Talk

Our look was as if two lovers, or deadly enemies, met unexpectedly on an overgrown path when each had been thinking of something else: a clearing blow to the gut. It was also a bright blow to the brain, or a sudden beating of brains, with all the charge and intimate grate of rubbed balloons. It emptied our lungs. It felled the forest, moved the fields, and drained the pond; the world dismantled and tumbled into that black hole of eyes. If you and I looked at each other that way, our skulls would split and drop to our shoulders. But we don't. We keep our skulls. So.

He disappeared. This was only last week, and already I don't remember what shattered the enchantment. I think I blinked, I think I retrieved my brain from the weasel's brain, and tried to memorize what I was seeing, and the weasel felt the yank of separation, the careening splashdown into real life and the urgent current of instinct. He vanished under the wild rose. I waited motionless, my mind suddenly full of data and my spirit with pleadings, but he didn't return.

Please do not tell me about "approach-avoidance conflicts." I tell you I've been in that weasel's brain for sixty seconds, and he was in mine. Brains are private places, muttering through unique and secret tapes—but the weasel and I both plugged into another tape simultaneously, for a sweet and shocking time. Can I help it if it was blank?

What goes on in his brain the rest of the time? What does a weasel think about? He won't say. His journal is tracks in clay, a spray of feathers, mouse blood and bone: uncollected, unconnected, loose-leaf, and blown.

I would like to learn, or remember, how to live. I come to Hollins Pond not so much to learn how to live as, frankly, to forget about it. That is, I don't think I can learn from a wild animal how to live in particular—shall I suck warm blood, hold my tail high, walk with my footprints precisely over the prints of my hand?—but I might learn something of mindlessness, something of the purity of living in the physical senses and the dignity of living without bias or motive. The weasel lives in necessity and we live in choice, hating necessity and dying at the last ignobly in its talons. I would like to live as I should, as the weasel lives as he should. And I suspect that for me the way is like the weasel's: open to time and death painlessly, noticing everything, remembering nothing, choosing the given with a fierce and pointed will.

I missed my chance. I should have gone for the throat. I should have lunged for that streak of white under the weasel's chin and held on, held on through mud and into the wild rose, held on for a dearer life. We could live under the wild rose wild as weasels, mute and uncomprehending. I could very calmly go wild. I could live two days in the den, curled leaning on mouse fur, sniffing bird bones, blinking, licking, breathing musk, my hair tangled in the roots of grasses. Down is a good place to go, where the mind is single. Down is out, out of your ever-loving mind and back to your careless senses.

—5—

The Way to the Forest

A Story from Chernobyl

Joanna Macy

I t is hard to renew the spirit if its ancient connection with earth has been severed, and hard to work for the healing of the natural world when we are wrenched away from our place within it. Yet, as I learned with the people of Chernobyl, these once-powerful connections can still be tapped. They seem to be within us.

Each year, each day now, more humans are uprooted from the lands that held the memories of their people. Some of us can go back to visit the old country places, but we are likely to find them carved up by freeways, industrial parks, and shopping malls. Countless millions of others can never go back. Driven out by political upheaval, persecution, civil war, they crowd into refugee camps or wander the slums of alien cities far from the fields and forests of home. Some are exiled by economic forces, their ancestral holdings swallowed up by distant firms for monoculture of export crops, while others are hunger-driven from homelands that drought and erosion have turned to desert. Other environmental refugees leave because, through industrial disaster, their portion of earth has become too contaminated to support life.

And there are those who are unable to leave their contaminated homeland. They remain—but removed from nature, and fearful of it. They live in internal exile.

After the explosion in late April 1986 that turned Chernobyl's fourth reactor into a volcano of radioactivity, many inhabitants of the surrounding areas left. Thousands of families were evacuated to less contaminated localities, sometimes moving again, and again, as the radioactivity spread. Many more thousands did not leave. By choice or default they stay behind.

Many remain still waiting for jobs and housing to move to, or still weighing the decision to move at all. They see those who had evacuated earlier begin to trickle back, choosing home—even a contaminated home—over the bleak anonymity of exile. The new life is no life, say those who return, when your family and relatives are scattered, when you are stranded among strangers in a faceless highrise, when out of ignorance and fear your children are teased and called "Chernobyl hedgehogs" and "mutants."

Meanwhile the radiation from Chernobyl spreads. With each passing year since the disaster began, its consequences proliferate. Radioactive isotopes, carried by wind and water, contaminate new areas, create new toxins as they mingle with industrial pollution. When radioactivity from the disaster is found in fresh food as far away as Minsk and Rostov, it can seem futile to try to escape, and easy to imagine that nothing is really clean any more, that nowhere is really safe.

So, in the more toxic areas closer to Chernobyl, many simply stay put. And staying put, they try to stay clear—of the fields, forests, streams around their homes. "Don't sit on the grass! Don't wander in the woods," they admonish their children. "No playing in the stream, no fishing, no swimming, no tree-climbing, no berry-picking. . . . Stay inside."

It was in just such an area that I found myself in the fall of 1992, with a Russian-American team of mental health workers. We had come to Novozybkov.

"Our aim," said the mayor of Novozybkov, "is to get people off the ground, away from the soil, the trees; so we plan to construct more apartment buildings." Novozybkov had been a thriving agricultural and light industrial center town of some 50,000 in the Bryansk region of southwestern Russia. About a hundred miles from Chernobyl, it is probably the most contaminated town of its size that is still inhabited.

"Over half our population still live in the old wooden houses, close to the earth," the mayor explained, referring to the traditional, single-story bungalows. Trimmed with color and carving, and nestled in greenery, these dwellings look like illustrations of Russian folktales. "As soon as sufficient cement is procured, we will be building more six- and eight-story blocks of flats, so that we can demolish the old houses. Made of wood, heated by wood, they must go, for wood holds the radioactivity."

The mayor can consider such large scale measures now that the fate of the city has been finally, officially determined. In early 1992 Boris Yeltsin came in person and with fanfare proclaimed: Novozybkov will stay. Novozybkov will live. For the six preceding years, since the Chernobyl disaster began, city and county government had been virtually immobilized, suspended in a kind of limbo. The local authorities neither evacuated the inhabitants nor created new facilities and programs for them. Their apparent ambivalence contrasted with the more active response eventually adopted by officials in nearby Ukraine and Byelorussia, or Belarus as it is now named, and probably relates to the particular way in which their area became so heavily contaminated.

On April 27, 1986, the first winds carrying fall-out from the burning reactor blew north-northwest over Byelorussia and the Baltic states, setting off in Sweden the alarms that alerted the outside world. As the winds shifted, they carried the billowing radioactivity to the West in a broad swath over the Ukraine, Czechoslovakia, Germany. Yet another wind-shift blew smoke and clouds northeastward in the direction of Moscow and its huge, heavily populated metropolitan area. But the capital area was saved, because en route, over southwestern Russia, the clouds opened—and a heavy, late April rain bearing radioactive iodine, strontium, cesium, and particles of plutonium drenched the towns and fields and forests of the Bryansk region.

Years later in Moscow I would be told that those clouds had been seeded; but at the time it was not the kind of political decision that is officially announced. Even six years later when it seemed to be common knowledge, I heard little mention of it, as if the decision deliberately to sacrifice one area in order to protect another were no more grievous than the Chernobyl explosion itself.

Our Russian-American team of four, which consisted of a psychologist and a physician from Moscow, my Russian-speaking husband, and myself as team leader, were calling on the mayor to acquaint him with the workshop we had come to Novozybkov to conduct. We had just arrived from working with Chernobyl victims in Belarus. The purpose of our workshops, we explained to the mayor, was to help people understand and respond creatively to the psychological and social effects of massive, collective trauma, so that they could overcome apathy and divisiveness and develop mutual trust and skills for self-help.

Our work drew on insights and methods which had evolved in the West over the last fifteen years, becoming known as the "inner" work of social action and as "despair and empowerment work." Offering psychological and spiritual resources for facing issues of collective survival, it had proved useful in building trust and courage for mutual support and creative action. We had come to Novozybkov to share some of these methods, in the hope that they would be useful here as well, in the shadow of Chernobyl. We reported to the mayor that, in response to the announcement, over a hundred citizens of Novozybkov had come forward, eager to take part in the three-day event. Unfortunately, due to lack of space and the nature of the work, only half that number could be accommodated this first time.

The mayor, a handsome, heavyset man of about forty, listened guardedly. "It is good of you to come to undertake psychological rehabilitation," he said. "There is, however, a team from Moscow Pedagogical University that is planning to come in November and begin an extensive program of psychological rehabilitation." He was clearly skeptical about what we, as outsiders, could contribute in three days.

I knew that "psychological rehabilitation" was now the preferred term for dealing with the trauma of Chernobyl. Although glad that the mental and emotional costs of the disaster were acknowledged, I considered the term an affront to the people. It reduced their suffering to a pathology, as if their anguish for their families and their homeland were not a normal response to the cataclysm, and as if it could be corrected, fixed. The people of Chernobyl had already suffered a lot from efforts to psychologize their suffering. In the first three years after the disaster began, doctors were ordered by the Ministry of Health to give no diagnosis mentioning radiation. Instead, when people insisted that their sickness and exhaustion, their cancers, miscarriages, and deformed babies had something to do with Chernobyl, they were diagnosed as afflicted with "radiophobia," an irrational fear of radiation.

I wondered if, in the few moments we had, I could convey to the mayor how our team's approach might differ from the notion of "psychological rehabilitation." Taking a deep breath, I said to him quietly, "We do not imagine we can take away the suffering of your people, which is very real. But we can look together at two main ways people respond to suffering. The suffering of a people can bring forth from them new strengths, great solidarity. Or it can breed isolation and conflict, turning us against each other. There is always a choice."

At my words the mayor's demeanor suddenly changed. Sighing and leaning back in his chair, he spread his hands on the table and said, "There is not a single day, not a single encounter in this office, that does not reveal the anger that is there just under the surface. Whatever the issue at hand, there is this anger that explodes or is barely contained, ready to explode."

Then he asked, after a pause, "Let me know if there is anything I can do to support your work here."

That afternoon, after visits to many classrooms with the school superintendent, I took tea in the home of my Novozybkov hosts and rested my eyes on a beautiful woodland scene. Our host family lives in a fourth floor flat in a cement block. The forest scene, sunlight flickering through birch trees into a grassy glade, is a sheet of wallpaper covering one side of the room. In the parlor crowded with overstuffed furniture it brings a refreshing sense of space. I commented on it to Vladimir Ilyich, the school superintendent, who was sitting there with his eleven-year old grandson and showing me the large Geiger counter he carries around in his car. He continues to measure the radioactivity, taking seriously his responsibility to the schools in the area in monitoring the environment for the children.

Following my eyes Vladimir Ilyich said, "You know, that is where we can't allow the children to go, or any of us for that matter. For the forests stay radioactive. This is hard, you see, because the forests were always important to us. Every weekend, every holiday, we'd be there, walking, picnicking, mushrooming. . . . They are our history. Our ancestors came from the forests. During the Nazi occupation, the partisans fought from the forests. Yes, we were always people of the forests." Quietly he repeated, "We were people of the forests."

I asked him, "When do you think your people will be able to go into the forests again?" With a tired little smile, he shrugged his shoulders. "Not in my lifetime," he said, and indicating his grandson, he added, "and not in his lifetime." Then he gestured to the wallpaper, "This is our forest now."

While we were talking over tea, the Russian members of our team were meeting with the people of Novozybkov who had shown up to register for the workshop. Their task was to sort them out, because over a hundred wanted to participate. We could have accommodated that number by having them sit in rows as in a theater, looking at the backs of each other's heads; but we wanted them to be able to see and address each other and have space to move around.

The next morning, the workshop began. In the auditorium of a school for special education, fifty people of Novozybkov, mostly teachers and parents, and mostly women, were seated with the team in a large circle. When invited to introduce themselves and to speak of their concerns, their earnestness was striking. They seemed to listen with bated breath, and they stood up to speak, the way their children stand in school when called on to recite.

A man said, "We can't keep waiting for someone else to rescue us. We only have ourselves, and we'd better get used to that notion." The references to Chernobyl were indirect, the name was not spoken. They referred instead to "the event." "Since the event," they would say. Nor for the first day and a half did they speak about the radioactive contamination of their town, their homes, their food, or about the illnesses they and their children experience. When a married couple took turns leaving every morning and afternoon, they said no word about their little daughter in the hospital, to whose bedside they were going. Their silence seemed to say, "This we don't need to talk about. We've got our nose in it all the time. We are here to see how we can achieve some kind of sanity and harmony in our family life."

In fact, that is how the workshop had been entitled in the local announcement: dealing with the family conflict and restoring family harmony. We had learned this upon our arrival in Novozybkov, and we were startled. Here we had come to help with the trauma of Chernobyl and it wasn't even mentioned. But our teammate, the Russian psychologist, reassured us. "Listen, they know it's about Chernobyl because the announcement says 'American Psychologists Here on a Humanitarian Mission.' They know perfectly well that humanitarian means in response to Chernobyl; you don't need to say the word. What's really bothering them on the conscious level is the tension and strife in their family relations."

In introducing the work we would do together, I drew a context larger than Chernobyl. I explained how in a number of other countries, including the United States, this kind of workshop helps people develop strength and courage in times of difficulty. "There are many people throughout the world who suffer greatly now from severe pollution and environmental catastrophes," I said and named a few, like Bhopal and Prince William Sound. "We can learn to let this suffering strengthen rather than weaken our families and communities, and bolster our will to act together for the healing of our world." Just as radiation attacks the physical immune system, it also, I pointed out, can weaken our cultural immune system, eroding our self-respect and our trust in each other. But there are deep wellsprings within us and our relationships that we can turn to and draw upon for strength and courage.

I sensed the eagerness of the people in the room to hear that their suffering was not isolated, but part of a global phenomenon of disregard for life, and integral to a larger trauma through which we all are passing now. So, as I spoke of what we all have to learn in this time—about telling the truth, about seeing clearly, about trusting each other, and taking risks together, I presented this challenge as one that we *all* confront.

During the rest of the day and half the next we offered structured processes to build trust and openness in communicating personal experience. Interactive exercises in pairs and small groups helped participants take risks in telling the truth about their thoughts and feelings, and also to discover their capacity to listen supportively without judgment or interruption. Family communications were the focus of another set of exercises, where participants played roles of parent and child in dialogue, and then let their own early experiences shed light on current dynamics in the home. Though such work was new to many, they engaged willingly, becoming increasingly spontaneous as they spoke and attentive as they listened. A matronly school psychologist expressed her enthusiasm. "This is a revolutionary experience," she rose to say; "for the first time in my life I have the undivided attention of a man!"

And we danced. I had recently learned a Lithuanian circle dance, whose movements were simple and stately. Called the Elm Dance, its theme was the healing of trees. The people of Novozybkov, who had all but forgotten their own traditional Russian dances, asked for it at the start and close of each day. As the music came on, all talk would stop and, taking each other's hands in concentric circles, their eyes half-closed in concentration or revery, they would move in unison the patterned steps.

It was not until the afternoon of the second day that the grief broke. It happened unexpectedly, at the close of a guided journey in which, moving backward through time, the participants were invited to remember their ancestors and harvest their strengths. As the exercise closes, they are invited to return to the present time, but in Novozybkov they did not want to. It felt good to recall earlier times, to evoke the countless generations who had lived on these lands and in these forests, and to imagine one could join them, but when we moved forward through time to 1986 and into the disaster years, the raw horror of Chernobyl could no longer be repressed.

Paper and colored pencils were at hand for capturing each person's image of the journey through time and the fits of the ancestors, but now these drawings had but a single theme. As the pictures were completed and shown to each other, talk exploded, unlocking memories of that still unacceptable spring—the searingly hot wind from the southwest, the white ash that fell from a clear sky, the children running and playing in it, the drenching rain that followed, the rumors, the fear. A number of the pictures featured trees, and a road to the trees, and across the road a barrier, or large *X*, blocking the way.

When we reassembled in the large circle, the feeling of mutual trust and enjoyment, that had been building since the workshop began, was shattered by anger. "Why did you do this to us?" a woman cried out vehemently. "What good does it do? I'd be willing to feel the sorrow—all the sorrow in the world—if it could save my daughters from cancer. Each time I look at them I wonder if tumors will grow in their little bodies. Can my tears protect them? What good are my tears if they can't?"

Angry, puzzled statements were made by others as well. Our time together had been so good until now, so welcome a respite from what their lives had become; why had I spoiled it? I listened to them all, wondering what I could possibly say. To sermonize about the healing potential of their tears, or give clinical explanations about the value of despair-work, was out of the question. It surprised me that when I finally broke the silence that followed their outburst, I spoke not about them or Chernobyl, but about the people I had worked with in Germany.

"After the war which almost destroyed their country, the German people determined that they would do anything to spare their children the suffering they had known. They worked hard to provide them a safe, rich life; they created an economic miracle. They gave their children everything—everything except for one thing. They did not give them their broken heart. And their children have never forgiven them."

The next morning as we took our seats after the Elm Dance, I was relieved to see that all fifty were still there. Behind us, still taped to the walls, hung the drawings of the previous afternoon, the sketches of the trees and the slashing X's that barred the way to the trees. "It was hard yesterday," I said. "How is it with you now?"

The first to rise was the woman who had expressed the greatest anger, the mother of the two daughters. "I hardly slept. It feels like my heart is breaking open. Maybe it will keep breaking again and again every day, I don't know. But somehow—I can't explain—it feels right. This breaking connects me to everything and everyone, as if we were all branches of the same tree."

Of the others who spoke after her that last morning, the one I recall most clearly was the man I recognized as the father who regularly stepped out to visit his little girl in the hospital. This was the first time he had addressed the whole group, and his bearing was so stolid, his face as expressionless as ever. "Yes, it was hard yesterday," he said. "Hard to look at the pain, hard to feel it, hard to speak it. But the way it feels today is like being clean, for the first time in a long time." The word he used for clean (*chisti*) is also the term for uncontaminated.

For my own sharing to the group I spoke of the meeting I would attend the following week in Austria, the World Uranium Hearing where native peoples from around the world would testify to their experiences of nuclear contamination. "They are many," I said, "and they are speaking out about the disease and death that follow in the wake of nuclear power and weapons production, at every step from mining to testing. You are not alone in your suffering. You are part of a vast web of brothers and sisters, determined to use their painful experience to help restore the health of our world. Feel your solidarity with them, as they will with you; for I will speak of Novozybkov and tell your story there at the Hearing and to my own people back home."

The following hours were devoted to planning, because the group wanted to continue, after the team left, the kind of work we had been doing and to bring it to others in the area. When in the next week could the group meet again, and where? Who would serve as communications link? Who would locate a room? Copies were made of the music we had used; the Elm Dance especially was in great demand.

At the end we played it one last time. It felt good to let words subside as we formed into concentric circles and moved into the slow, fluid, and familiar steps. The mute interlacing of hands, the swaying of bodies in rhythm, seemed more eloquent than speech.

The early evening was still bright when our team boarded the overnight train to Moscow and hung out of doors and windows to shout last good-byes. For the next hour, until it was too dark to see, I stood at the open window watching for the stretches of forest, peering out into the trees as we passed. I wanted to behold them more distinctly than the dusk and the speed of the train permitted. I wanted to see what I had already glimpsed of them in the people of Novozybkov, to see how firmly they were rooted and how their swaying branches interlaced.

—6—
A Letter for Montana Wilderness

Rick Bass

Mike Espy
U.S. Representative, Mississippi
Dept. Of Agriculture
U.S. House
Washington, DC 20515

Bruce Babbitt
Secretary of the Interior
1849 C Street NW
Washington, DC 20515

Dear Mike Espy and Bruce Babbitt:

I lived in Mississippi from 1980 to 1987, where I worked as a petroleum geologist, and was active in the Central Chapter of the Sierra Club. I remember voting for you, Representative Espy, and being elated, and surprised, when you won. I have been impressed with your work in Washington, for Mississippi and all Americans, ever since. Secretary Babbitt, I was likewise thrilled with your appointment, and am hopeful you'll be allowed to help start reclaiming the public lands for the public.

I am living in northwestern Montana now, in a small valley called Yaak. It's 97 percent National Forest lands, under the jurisdiction, of course, of the U.S. Department of Agriculture. This week we had some friends visiting us from Mississippi who expressed dismay (a polite phrase) at the clear-cuts on the upper portion of this remote valley, the Yaak Valley. The clear-cuts were up on the northernmost border, up where no one ever went. Hundreds of acres in size, the clear-cuts are on slopes so steep that the soil was washed away, leaving bare rock, where nothing can grow. Over twenty years after the logging, entire mountain faces are still totally bare, and will probably not ever revegetate. But the local timber industry got their cut out, and moved on.

It is a wild valley still, because it is on the U.S./Canadian line. It's lush and wet, rich in biodiversity, where it hasn't been clear-cut. It links the ecosystems of the northern Rockies to the Pacific Northeast—in that respect, it is the most strategic location in the West, with regard to the concept of conservation biology, and the flow and exchange of wild genes. It is home to grizzlies, wolves, wolverines, a woodland caribou, great gray owls, bull trout, sturgeon, and a host of other threatened and endangered species, and yet not *one acre* of designated federal wilderness exists in the Yaak.

Part of this is because it is not a "pretty" place for people to go and admire scenic vistas of ice-capped peaks; the forests that remain are either too thick or too fragmented (over ten thousand miles of roads, built into these last roadless areas at taxpayers' expense). These roads went first and fastest into the biggest, oldest forests, cutting the largest, most valuable trees for the international corporations. What's left isn't pretty, but it's wild; it's wilderness. It's where wild things live.

It rains a lot, in Yaak. The very thing that keeps people away is what helps make this valley, despite it historical ravages from the local timber industry, a safe harbor for these wild, shy creatures. It's low elevation rain forest; it grows big trees.

The reason Yaak has not a single acre of designated wilderness is the power of corporate politics. Ten thousand people in Lincoln County used to vote one way: pro-mill. This is a small number of people and a small number of Montanans, but it's enough to have been perceived as a swing vote in Congressional politics, and some politicians in the West have a bad habit of controlling completely that public domain to which they are privy, and from which their constituents are so fortunate to receive the many benefits. It is my valley, my home, which the local mill, run by Champion International, has held hostage from the rest of the country for so many years.

But not any longer. I wish I could say it is because Congress has decided to protect the last of this country's federal wildlands. However, that's not the reason. The reason is that Champion, having cut almost all the forests, is leaving town—shifting operations down toward Mississippi, Arkansas, and other southern states, where the trees grow faster. They're coming your way, Representative Espy; be warned. They are not interested in preserving communities, or preserving anything.

The reason they're leaving is that in the late 1980s, to raise cash and reduce assets in order to defend against corporate takeovers that would pirate and fragment the company (in the same manner in which they had fragmented the land), Champion liquidated their assets as a defensive maneuver, to raise cash. Their assets, unfortunately, happened to be whole forests—not pine monocultures or "fiber factories," but complete riparian forests of larch, spruce, fir, and pine. In all instances, Champion's lands were adjacent to federal lands, and the wild creatures that moved through and among and above the forests did not understand, I'm sure, the murky betrayals of property boundaries. They only knew the land—their community—as a whole.

Over the course of only three years, Champion clear-cut almost all of their properties—867,000 acres in the state of Montana—most of it river bottom lands, the most critical wildlife habitat zone that exists. But before they wiped out all these forests, Champion made sure they were covered; they cut a deal with Congress, and with the U.S. Fish & Wildlife Service, that would guarantee that Champion could continue cutting ad infinitum at this artificially high, hyperinflated volume; they obtained assurance from the Forest Service that after Champion ran out of all its logs, it could continue at the same pace on the federal lands. Never mind the damage to the ecosystem, or to the various species within that system; Champion wanted subsidizing, wanted exception from environmental regulations concerning endangered species, and wanted to keep, solely for profit's sake, the same steady (and too-high) volume running through the mill.

Because the Champion-owned mill in Libby would benefit in the short term by this increased, unsustainable volume of harvest, Champion presented themselves to the town of Libby, and Lincoln County, as purveyors of goodness, the cornerstone of civic responsibility: creators of jobs. (When actually they were robbing jobs by clear-cutting rather than the more labor-intensive selective cutting, and when they were sending raw logs from federal lands over to Asia, where the logs would be milled and manufactured and then sold back to American consumers as finished products at ten times the price at which they'd left the country. But because the taxpayers were paying, and because Champion was getting the free ride, it was all about volume and speed back then, not quality or sustainability. Champion knew all along they'd be leaving soon. . . .)

Environmentalists warned Libby that this was what Champion was up to, that Champion would screw the community the way they had screwed the land, but the political force of a small-town company town is well-known; the environmentalists were labeled (by Champion) as anti-growth and anti-American, and I regret to say that there is now yet one more instance of a boom-town gone bust. When I was a geologist in Mississippi I saw the price of oil plunge from $40 a barrel to below $10 a barrel, true boom-and-bust, but still even that is nothing compared to the way Champion went through the federal forests of Yaak before fleeing.

Watching it all has confirmed for me something I've always believed: that respect cannot be turned on and off, like electricity. Either you've learned to practice respect, or you haven't. I have been thinking that many of our problems in the American society stem from how we began, during the agricultural revolution, to use slaves—the voiceless minorities—to control the land, and from there leapt hot-headed and wrong-hearted, already on the wrong course, or with false hearts, into the industrial revolution. Machines allowed us to treat the land with a newfound, careless lack of respect for forests and rivers and wild things, and this metastasized further into a lack of respect for all other voiceless minorities. Slavery was ended, but only on paper; we still had not learned respect. Or rather, had forgotten it.

In that sense—our culture, our heritage of dominating the voiceless minorities, the silent partners—I think that all of our problems are in one sense crises of the environment. I think that there is no separation, at heart's depth, between pollution, large-scale clear-cutting, homelessness, illiteracy, drug abuse, or poverty. They are all a crisis of disrespect. And I believe it is an epidemic: north, south, east, and west.

Right-wing critics have long sought to divide and conquer those who would argue for respect by attempting to pit environmentalists against other social activists, accusing the environmentalists of loving animals and forests more than people. (And in the meantime, the chainsaws rev louder, the bulldozers accelerate, and the mountains slide away into the muddy streams as the land developers, the land scrapers, call in their political favors and swing their three- or five-year projects onto the public lands, reap their profits, grow old, grow fat, and then die, fading from history and memory like the dying glow of a forgotten firefly, completely undistinguished in the annals of contributions to the human spirit and human dignity, instead leaving behind an anonymous legacy of ruined land and lost potential.)

We can't teach these kinds of people how to have respect for the voiceless; we can't muscle respect down their throats.

But if we can preserve some intact systems in the world where respect and logic still function in harmony—where actions still have consequences—and if people and communities can observe nature's cycles, both minute and grand, of respect, cooperation, and compromise—and if people can grow up learning to accept that there must be a few last wild places into which our furious domination will stop, a few last places into which we will not bull our way for profit—then I believe there is hope for other of our culture's ills of domination and loss of accountability. I believe that if we can learn to give up a tiny bit of power, only then can we learn to be strong. We were strong in our nation's beginnings because we were young, and because our country was blessed with an excess of natural resources—and now we must learn to be strong because we are old: to be wise, and not follow the bitter road to the very end of the last of those resources, now that they are so rare and precious—now that we have gone beyond sustainability.

I see these several years, on toward the end of the century, as the last chance we have to re-learn respect: for the land we live on, and respect for ourselves, and out of that, respect for all others. It may be romantic or naive but I am convinced that if a person can watch even a brief glimpse of unfragmented nature and all its woven connections—the way deer fawns are born the same week in early summer, for instance, that the grass is highest and greenest, to protect them from predators—then we can still hope to re-learn the forgotten notions of responsibility and accountability; of respect. In nature first and clearest we might be able to re-learn the oldest truth that everything is, and always has been connected; that a child's poverty in Mobile, Alabama, for example, robs from the security and strength of all of us.

The land got sick first, as a result of our dominating the voiceless minorities, and it is the land which must get well first, I think. I do not believe we can bypass our relationship to the land if we hope also to improve this culture's relationship to women, to children, to people of color, to the poor, the illiterate, the homeless. . . . We can't be good to each other if we can't even be good to a tree, or a forest.

It is not a question of priorities, as our foes would accuse us; it is not a question of jobs versus wilderness, or wild nature versus education for the hungry child in Mobile, or Portland, or Philadelphia. It is a question of finding strength once more—strength in wisdom and restraint, this time, rather than the fast-diminishing strength of aging muscle, or calamitous, misdirected passion.

Let me return to my valley, please—the Yaak. Let me leave the abstract and return to the specific: this wild rainy place where there are not vistas, only clear-cuts and a few wild shy creatures, such as the wolves and myself, living up here in a magical seam between the ecosystems of the flora-rich Pacific Northwest (*golden spleenwort, linear-leaved sundew, Queen's-cup bead lily*) and the magnificent predator-prey showcases of the northern Rockies: bear-elk, wolf-moose, cougar-deer, owl-grouse, eagle-trout, lynx-hare. . . .

Like no other place in the world, it is all still here, and all still connected, though just barely. We're down to five or six wolves, in this valley of almost 300,000 acres; down to nine or ten grizzlies. One caribou, occasionally, drifting back and forth among the lost no-man's lands between northern Idaho, Montana, and Canada. Twenty bull trout, and a handful of aging white sturgeon in the mighty Kootenai River.

I am not asking for tourists to flock to this wet, cold valley; unlike our other, prettier, rock-and-ice wildernesses, this one is not for humans. What I'm asking is for action by Congress to help save the last hidden voiceless things up here.

Once again, the Yaak wilderness is not a wilderness for people to visit, or to attempt to thrash their way through the wet dripping ferns. They're not things to be seen, even if you could, which you can't—these few wintery mountaintop wolverines that drift back and forth across the Canadian line. What the greatest value this valley has to offer us is the way we choose to protect it—not the way we choose to tour or develop it. As many of us did for Alaska's wilderness in the 1980s, Yaak is a place to fight for until the fight is won, and then to know simply that here is a wild place we don't need to enter, that's not designed for our entrance—brushy jungle—a place where we demonstrated our strength by the manner in which we chose to give up our compulsive, chronic power over it, while the country still meant something: while there was still a bit of sweetness left to it.

I wrote this damn book about this valley, some time ago, back when I was naive; I referred to the valley by its proper name, Yaak, rather than some made-up name of fantasy. For whatever reasons, the book struck chords in the hearts of readers, of which there have been many. It was a story about starting a new life, and noticing new things, in a new place.

I've been up here seven years. It used to be close to Eden; up until I wrote that book, there was rarely a car on the road. People rode horses down the road. You could walk down the road and see a mountain lion napping in the dust in a patch of sun. Grouse, too, would come out and dust-bathe along the road, or peck gravel for their gizzards; sometimes a hawk would plunge from the sky and nail one, right before your eyes. If a snowshoe hare ran out in front of you, you paused half-a-second; chances were good that a coyote, or bobcat, or even a wolf would be hot on its heels. Badgers sunned themselves dry on the roads after late-day summer thunderstorms had filled their holes briefly with water. Bears padded across the roads, quickly, furtively.

Since my book, all that's changed. The animals are still out there, but they're wary, because there are so many more cars now. A good number of the drivers of those cars come searching for me, uninvited, to knock on my door and interrupt the living dream of my life, the waking dreamtime, to tell me they read the book and liked it (or didn't). That's my problem, and my due, I suppose. And perhaps it's best that the animals have learned to move away from the roads; perhaps this increased wariness will serve them better, in the future. Still, I feel a sense of loss, of unnecessary, lost innocence.

This in not to say that there are not still infrequent surprises from out of the blue, when the animals reveal themselves. A great blue heron spearing tiny brook trout; a golden eagle among thirty ravens, sitting atop a road-killed deer. One winter morning my wife and baby daughter and I looked out the window at the pond to watch a family of five otters playing on the pond ice at twenty below, diving in and out of a hole in the ice, catching fish and playing tag. Yesterday evening, a friend drove up to visit, and as we sat there talking until dusk, a badger crawled under his car, perhaps to lick antifreeze from beneath his slightly-leaking radiator—his car had overheated while chasing elk poachers earlier in the afternoon—and the badger charged us and my dogs (the same dogs that had been torn up by a pack of coyotes several weeks earlier). The badger then retreated to beneath the car and wouldn't leave, and wouldn't let Jesse or myself near the car. It growled and hissed and snarled whenever we got within twenty yards of it, and we had to sit on the porch and wait until after dark before the badger waddled away, flowing silver across the yard under a perfect full moon. . . .

What is the point? The point is that here is a minority—the minority of the wild, the minority of the untamed, the uncompromising—and that it is in our power to give back its freedom—to allow the last two roadless areas—Roderick/Grizzly, and Mt. Henry/Pink Mountain/Gold Hill—to remain what they are, and to become what they will, under their own design, by their own processes. Generosity, Barry Lopez has noted, is an act of courage. Protecting one of the last two areas is not enough; we need to protect both of them for there to be a weave, a relationship of the wild.

The valley to the east of here, the Ural Valley, once home to grizzlies and wolves, has disappeared completely—was flooded by the building of Libby Dam, which generates electricity and sends it to Tacoma, which then sells it via the Bonneville Power Administration directly to California.

Seventy miles to the south, the world's largest copper and silver mine is proposed to tunnel beneath the Cabinet Mountains Wilderness. The mining company (Canadian-owned Noranda), the U.S. Forest Service, and the U.S. Fish & Wildlife Service have broken up time-honored family home ranges of grizzly bears, the historic and ancient cultures of grizzly families, into arbitrarily-numbered compartments called Grizzly Bear Habitat Management Units—GBHMU's—and like little children bullying something smaller or more voiceless than themselves, the agencies claim to always have habitat available to the bears, *somewhere*, at a 70 percent habitat effectiveness rating. (When I went to school, 70 percent was a D-minus.)

Furthermore, the agencies shuttle these effectiveness ratings around every three years, via logging sales, and expect that these bear families, these bear cultures, will find their way from Compartment 1 to Compartment 37, and then when Compartment 37 is logged, on to Compartment 62. . . . The Forest Service calls this "displacement habitat," and feels clever, that they'd "provided" this displacement area (usually at that 70 percent level). It reminds me eerily of our government's shuttling of Native Americans from reservation to reservation.

Grizzly cubs stay with their mother for at least three years, so in effect, there will be grizzlies that are asked to move every time they become pregnant, and even worse, subadult bears will be expected to find their way from piss-poor Compartment 1 to piss-poor Compartment 37 all on their own, in their first wobbly year of independence. This is pseudo-biology, computer manipulation used to continue stealing from the public lands; it hides from the U.S. Congress and the taxpaying public the biological tenet that grizzlies live, or try to live, their entire life in a single home range. They pass on information about that one place—their *home*—to their off-spring: where to find berries, when to hibernate, where to hibernate, where to hide. . . . There's no word for it other than *culture*, and the Forest Service is pretending this doesn't happen. This is a deception of Congress and the American public; the culture of bears is being destroyed. Activities on the National Forests are destroying the bears themselves, just as killing all the bison a hundred-plus years ago finished off the culture of Native Americans. If the last two roadless cores in Yaak aren't protected, the grizzlies (and a whole lot of other species) aren't going to survive. It's simply not going to happen. We might as well just shoot them in the head and get it over with.

In this wild, unpeopled valley far from your homes, up on the Canadian/Idaho/Montana border, lies the clean but disappearing chance to keep a balance, and to preserve a working model, an example of a system that's still based on logic and the cooperative integration of every single element in that system, large or small, voiceless or bold. This, I believe, is nature's definition for the word *compassion*, which I'm afraid is a thing we move further and further from yearly.

I do not mean to sound like a salesman, as if I'm recommending preservation of wilderness for the sake of slowing our descent down the wrong turn that our culture (like so many ancient, extinct cultures before us) has taken. I believe it's much simpler that that: that we should preserve our last tiny remnants of this northernmost wilderness in the Lower 48—not just Roderick Mountain, but Grizzly Peak and Mt. Henry, Pink Mountain and Gold Hill, in this cornermost, forgotten valley—because it is simply the right thing to do.

These days we no longer have the comfort of hiding out in a moral middle-ground, a no-man's land of no-decisions. These days—down to our last one percent of unroaded American wilderness—if you are not acting to save it, then you are aiding in its development. Every one of us must line up either on the side of compassion, or on the side of manipulation, control, and domination.

I refuse to believe these latter ugly things are hard-wired into our existence. I continue to hope and believe we have within us the ability, the grace, to finally learn a light touch—to learn to lift our hands from a thing and step back and allow it to take its own course, to choose its own way, at its own pace. If we can't do that on such a small area as these 55,000 acres, the last we have, what chance do we have of a similar grace returning to and sweeping across the whole country?

Let anyone who reads this plea please act to save a long-ignored, long-abused voiceless place: to write letters for these last Yaak mountains without ceasing. Protecting Roderick alone as wilderness is not enough, anymore than being compassionate half the time is enough; this would only create an island, rather than protect a woven, breathing system of grace. We need to protect this last *system*.

As ever, there will be some who will oppose the notion of protecting the last wilderness we have—a thing that won't ever be coming back. There will be a few who will want to build roads into and through and over these last four mountains in Yaak, after having warmed up on the first ninety-six. Even if you don't recognize these people at first, you will later. They're the ones who aren't planning to stick around. They're the ones who are in a hurry, on their quick way to exercise their unlearned dominance over the next voiceless population, and they certainly will not stop at trees and animals. You will recognize them after they have gone on into history, and have taken with them a thing that might otherwise have remained beautiful, whole, and that rarest of things, free.

Thank you for your time—for listening.

Sincerely,

Rick Bass

cc: George Miller, Chairman, House Committee on Natural Resources,
   National Parks, Forests, and Public Lands Subcommittee, and
   Bruce Vento, Subcommittee Chairman
   U.S. House
   Washington, DC 20515

cc: J. Bennett Johnston, Chairman, Senate Energy and Natural Resources
   Committee, Publics Lands, National Parks, and Forests Subcommittee, and
   Dale Bumpers, Subcommittee Chairman
   U.S. Senate
   Washington, DC 20510

Jack Ward Thomas	Regional Forester
U.S. Forest Service	Region One, U.S. Forest Service
Box 96090	Box 7669
Washington, DC 20090	Missoula, MT 59807

Max Baucus, D-Montana	Conrad Burns, R-Montana	Pat Williams, D-Montana
U.S. Senate	U.S. Senate	U.S. House
Washington, DC 20510	Washington, DC 20510	Washington, DC 20515

Also in the House (Washington, DC 20515): Subcommittee members Ed Markey (MA), Nick Rahall (WV), Peter DeFazio (OR), Tim Johnson (SD), Neil Abercrombie (HI), Carlos Romero-Barcel (PR), Karan English (AZ), Karen Shepherd (UT), Maurice Hinchey (NY), Robert Underwood (GU), Austin Murphy (PA), Bill Richardson (NM), Patsy Mink (HI), Jim Hansen (UT), Bob Smith (OR), Craig Thomas (WY), John Duncan (TN), Joel Hefley (CO), John Doolittle (CA), Richard Baker (LA), Ken Calvert (CA), Jay Dickey (AR), Gerry Studds (MA), Kika de la Garza (TX), Carolyn Maloney (NY).

Also in the Senate (Washington, DC 20510): Subcommittee members Bill Bradley (NJ), Jeff Bingaman (NM), Daniel Akaka (HI), Richard Shelby (AL), Paul Wellstone (MN), Ben Nighthorse Campbell (CO), Mark Hatfield (OR), Pete Domenici (NM), Frank Murkowski (AK), Larry Craig (ID), Robert Bennett (UT), Arlen Specter (PA), Trent Lott (MS).

President Bill Clinton	Vice President Al Gore
The White House	Office of the Vice-President
1600 Pennsylvania Ave.	Old Executive Office Building
Washington, DC 20500	Washington, DC 20515

—7—
The Idea of the North

David Rothenberg

The idea of the North marks a place for the idea of the wild, that latent possibility, lurking inside us, remaining untamed. Feared for centuries, this wilderness is at the edge of all maps, the country of the unknown. Cultures grow, expand, and conquer, based on the need to fill up these white spaces. Magellan made it around the world; suddenly there is less mystery.

But in the North, it is still cold. Few consent to its demands; only the intrepid, those needing the crisp silence to think and to exist, search for the North. Across the gray blue horizon, nothing is moving. Then a piece of white peeks out of the waves. It is nothing, it's harmless, just a bright glimpse. But beneath the dark surface is a mass of sharp danger. Still, we remember the light, the fullness of the sky.

This northerly want is a satisfaction within open spaces, an inexact science inviting our purest ideas: and home to a special aesthetic that has found at least one way to solve the problem of the wild.

What is the problem of the wild? Not the nostalgic yearning for a better past. Not quite. It's knowing that an affront to culture and order lurks always inside us, resisting all plans. The wild wants to break up our marriages, tear down our buildings, dissolve our corporations. It says no to our control. Yet without it we die. It keeps us wanting, and knowing the ambiguity of the path we must take.

The North answers this with clarity. To civilization, it reflects a cool indifference. Our skies are still greater than you. Our nights are alive with sparks and a glow that cannot be equaled.

There is the simplification of winter, of course. In reality it is a continuous soft twilight, punctuated by silent aurora dances in the hollow blue cloudless skies. It may be cold, but it is a crystalline chill, just enough light so you can see through it to forever.

But the idea of North is with us from any latitude, close by any season and surrounding heat. We need it to imagine more lucidity than the boom of usual confusion admits. It's wilder up there, fewer people, more lynx tracks, caribou, traces of bear. On his deathbed Thoreau murmured "moose . . . Indians" as final words in this life; he was not thinking of his trip to Cape

Cod, but of those long weeks in the interior of Maine, what he knew as the North, the immediate memory of wilderness always close to his heart even at his last moments on Earth.

Mark Helprin's wonderful novel *Winter's Tale* takes place in two divided times, one hundred years ago and today. In the old days everything was, as we often dream it, larger than life, with the winters colder than cold, snow billowing through the streets piling in drifts above the doors and windows above our heads. The characters travel from New York City north up the Hudson River on swift horse-drawn sleds over the ice, far into the interior northern regions of the state, to the mythical Lake of the Coheeries, where the tale of winter never ends, where the exhilaration, snow, and ice are permanent. Culture is not lacking, but is self-contained. The relatives living up there are brilliant at the kind of knowledge that can be sequestered and learned in isolation. There is one, I remember, who is a master of words, a linguist beyond compare, who has memorized dictionaries in those long winter nights and who finds no language puzzle to be in the least bit difficult.

Gary Snyder writes:

> The condition of life in the Far North still approaches the experience of the hunter-gatherer world, the kind of world that was not just the cradle but the young adulthood of humanity. The North still has a wild community, in most of its numbers, intact. There is a relatively small group of hardy individuals who live as hunters and foragers and who have learned to move with the mindful intensity that is basic to elder human experience. It is not the "frontier" but the last of the Pleistocene in all its glory of salmon, bear, caribou, deer, ducks and geese, whales and walruses, and moose. It will not, of course, last much longer.

Knud Rasmussen, the Danish explorer, alights on the coast of Thule, western Greenland, to investigate the northern country. The explorer is supposed to conquer, to give names to what outsiders do not know. But Rasmussen is different than most. He wants to hear what the people have to say, and ask for their own visions of their own history. He looks for anyone who looks like they might remember the stories of a lucid past, when the impossible may once have happened.

"Me? I'm just an ordinary old woman," says Nalungiaq, from under a thick fur coat. There is a photograph of her staring sternly from beneath the sealskin hood, gazing directly at the horizon, pondering the visitor's request:

"I haven't got much of anything to tell you." She hesitates a moment and then stares far across the snowscape and into the wind. "Well, I could tell you what life was like when the world began."

"Yes," says Knud, weathered but patient after years of searching for stories just like this one. "What was it like?" Her answer hints at how vast the power of human language might be, if this language might speak with the world, rather than define it away into silence. It might be the best story he ever found. Here are his remembered verses from the North:

> In the very earliest time,
> when both people and animals lived on earth,
> a person could become an animal, if she wanted,
> or an animal could be a human being.
> There was no difference:
> All spoke the same language.

And the dogs and wolves stop howling to one another for the moment as the words begin. People and animals living together on earth. That time could be now. And what is it that we are meant to share? Why do we care? Realizing our humanity seems to deny something animal that always walks with us. We watch them all moving shadowy around us. We imagine they can speak to each other, leaving us out of the talk. At least they seem to sense their own space in the world, knowing territory and purpose from birth all the way to death. We point to them running away from our recognition. They are, understandably, worried about us and what we might do to their world, immediately or far away at the limits of the North. We've lost our place because we are poised to wonder why. Yet once we knew one another's ways, and lived together in the identical community.

All is touched by the power of sound; people, earth, what lies in between. The visceral appeal need not have a reason underneath it. The flow of sounds has its own order, easy to love without knowing how it all works. This is one kind of magic: astonishment at happenings in the world without wanting to explain. The fire dies down, the song continues, and the words repeated again and again lose their senses and seep into sound:

> That was the time when words were like magic:
> A single word, spoken by chance,
> might have strange consequences.

Signifying syllables become the human mark, and here is a concise reminder of the greatest power language has: we say things, things come to pass.

Perhaps Nalungiaq means words are magic when they start events in motion without knowing the outcome. And that's the way it is, now or then, before or after, past to future. There is no control, with messages sent one way, received another, cries turning into calls into melodies into changes in the land and time, no action without reaction, no plan without a void unchanged, the unknown preserved.

"Taoq! Taoq! Taoq!" says the fox, "Dark! Dark! Dark!," so he can hunt for his prey without discovery. "Light! Light! Light!" sings the rabbit, "Uvdloq! Uvdloq! Uvdloq!" so he can bolt away without being captured. One single word was once enough to make a difference. Call those few words to life, and they blur the sense between the human and the animal, crossing the boundaries of what each species alone might understand. Just speak, and what you want may come true. That makes the time which is the first time real.

> It would suddenly come alive
> and what people wanted to happen
> could happen—
> all you had to do was say it.
> Nobody could explain it:
> That's the way it was.

A sad longing for a time we never did understand. Naming takes us far from the animal powers even as it brings us close. That's the way it still is. What is most important cannot yet be explained.

Rasmussen takes his time searching the coasts of Thule for the most magic of words. He scours this northern edge of the sea, sailing in and out of every bay. This is his fifth time on these nominally Danish but very foreign shores. It is 1925. He has long since ceased imagining any practical passage to ease navigation through the Arctic. By this time he has seen a tragedy rise at night with the northern lights: these are people with ancient stories that are fading. The value of the ebbing words will be worth more to outsiders than the practical resources of the land itself. The people he finds implore him to listen closely to the words, and to repeat them over and over again. These turn out to be the chains of syllables whose meaning no one can quite remember, with rumblings reaching out to the animal and spirit realms.

See the circle traced by a blade of grass blown by the breeze at the beach, reiterating a narrow arc in the sand. That too, part of the language. Receding water with the tide, ripples on earth as well as in ocean, no ridge exactly like any other. And on snowfields in the high mountains, melting each year unevenly, but in a way familiar, pockets and peaks called suncups, looking like the ruffled pattern on tidal sands, only in snow. An earth language, which you could feel in both places, running your hand across the wet surfaces. From speech to the silence of seasons moving over time. Magic words? We can lie quiet with them as the shapes of the earth talk back with shadows.

Glenn Gould, the pianist, decided early on he would not play for people, but live in his studio assembling versions of Bach and Mozart by splicing together a multitude of takes. The perfection of his music collages is legendary. Though once radical, his technique is now almost standard in the world of classical music today. What remains eccentric is his extremity, his drive to escape and live in a world composed only of music. What has his cutting and pasting to do with the North? "Living in the North," he hypothesized, never having been much beyond Toronto himself, "makes one into a philosopher." He tried also to articulate this feeling with a radio play, a collage of voices, stories fading in and out of life farther up toward the pole. I take my title from his, "The Idea of North."

Canadian composer R. Murray Schafer, following in the splice-steps for Gould, tried to explain what this philosophy of emptiness would mean for aesthetics: "The art of the North is the art of restraint. The art of the South is the art of excess." Here are lines from his story:

> I am a Northerner.
> My heart is pure.
> My mind is cool as an icebox.
> And the cold of the forest will be in me until my extinction.
> Between me and the North Pole may be a few dozen people.
> But I meet them rarely.
> Mostly I wander alone, making my peace with the northern lights.
> My head is a thousand acres of wilderness.
> At night my imagination howls with wolves.
> No matter how you cherish the wolf,
>> he will always look back to the forest.

Schafer's music heads closer to the wild with each year. He had a piece scored for dawn on a wilderness lake. Five thousand people showed up one morning to hear it. Then they went on to Vancouver late in the morning. Next he's writing for a week in the wilderness, an opera of participation only for those who are ready to give seven days of their time to submit to the work. The cold takes time. It is not for everyone.

Climbing the ridges up in Lofoten, mountains touching the ocean, unsung, too far North, unknown to the climbers far away. Torngat. Stettind. Svalbard. The crispness is not isolation, but precision.

> Start from the view:
>   from blue half-islands that encode a sea,
>   orange, perhaps, only cyan
>   bands where
>   sky equals cloud
>   turn grayless gray
>   wisped scree cliffs
>   mass of water and air
>
> —just the Norths, nothing more—

The story begins with the log of a ship trapped in the ice. But it has travelled all over the world to get there. Dr. Frankenstein and his monster have pursued each other to exhaustion and the final showdown is out there on the ice near the Pole. Few people remember this part of Mary Shelley's story. *Frankenstein* is an extreme tale, of the forbidden knowledge of the creation of life. The doctor has made an incomplete monster, he runs the globe amok in rage. They finally confront each other, leaping from one ice floe to the next. They will consume each other, the creator and his creature. Life conjured up out of death, a miracle gone awry. This wonder, this horror, confronted in the place coldest and most sublime to the Romantic imagination. The North, a place only the extreme ones of us will reach.

The North now means more than it once did, newly opposed to the image of the South. What used to be a duality between the first and the third, the developed and the developing worlds, is now the polarity North vs. South. We of privilege, power, and energy abuse are now all the North, and those in poverty, struggling, oppressed by our greed and progress are all the South. So much for all my spacious reveries. I retreat again into the story of a traveler. Tété-Michel Kpomassie recounts in his book *An African in Greenland* his childhood dream to reach those distant lands where all is cold and white. In his pilgrimage he seems to skip the polarity of the haves and have-nots for a greater pull of the compass. He writes of the Eskimo who endows every object with life, who senses souls everywhere. "In their eyes, all that drab, white, lifeless immensity of little intrinsic interest to an African like me—becomes a living world." But he learned to adjust, if not to blend in:

I adapted so well to Greenland that I believed nothing could stop me from spending the rest of my days there. . . . But if I were to live out my life in the Arctic, what use would it be to my fellow countrymen, to my native land? Was it not my duty to return to my brothers in Africa and become a storyteller of this glacial land of midnight sun and endless night? Should I not open my own continent to fresh horizons, and the outside world? That is why I decided to leave.

And we, too, all feel the pressure to return from the North, and the wild. Anything can be carried to extremes, but what if left there? I cannot connect these with other stories. They float like hidden blocks of ice in the dark swirling sea. I have told you only the tips of them, those parts that rise above the surface into the sky of my memories. The world now has too many stories for us to weave any whole truth around all. In the North there is space, room to breathe, some would say nothing to do because the pressures of the manacled world are so far from view. We remember only fragments, and these too recede quickly in the wake of the serious weight of the dark or the light. We can write down what we know, and tack it to the cabin walls. But soon these will freeze and be covered over with snow. We will run out of food. We will have to go home. Or learn to fish beneath the ice, to talk to seals and then kill them. And the distant troubles, far from our silent place, will also fall back and seem irrelevant.

Yet, I know I'll return to the mess and the crime and the crowds and the neighbors who live close together and smile when they see each other in the baking streets. The North remains my opposite, even as part of me remains there. Part of you is there too, even if you have never been. The antipodes of a circular journey, now I know that the North is not the wild. The wild is everywhere, always below the surface from the human point of view. It is North, it is South, East, and West. It surfaces when we accept that darkness and light make no sense without each other, when we see how each extreme can carry us away. But in most directions we miss what carries us, hidden beneath so much else we have made to cloak the original world with. In the North these things are stripped away. We are left with the immediate, immutable nature whose depths remain frozen to the human touch. We might mine or flood it, but its brightness does not fall.

Carry us up to exacting spaces, open around from the earth to the sky. And leave us there, not to die but to live.

Rest, Locate, Reckoning—

run a finger along the top of the map
that's where I am

still here:
        grabbing the frozen sand
        surprise and demand

"we shouldn't go there"
"no one should go"
—that's what you learn once you've been—

no use complaining, others will follow

watch the ships come in, glowing in the nightsun
        only outlines test the horizon

how else will you know these things?
how else will I find you?

Part II

—For the Love of All Animals—

—8—
Nighthawks

Linda Hasselstrom

On autumn nights, I always drive slowly down the dirt trail toward my house, because nighthawks often settle in the road; perhaps bare earth holds the day's heat. One evening, a nighthawk rose before the car and flew ahead of me. Each white spot made a half-arch as the wing rose and fell; each wing beat created a shining circle. As the car crept slowly forward, the nighthawk flew directly in front, wings flashing like beacons, light sparkling in a hoop about its dark body.

In the dark night, the nighthawk flies on within its own glimmering halo of pure grace.

Living in exile, I long for nighthawks.

I saw them that first evening as I helped my mother carry our suitcases to the tall house surrounded by green. But few of us *just* see nighthawks; if we are aware, they plummet like cannonballs into our lives.

The alfalfa plants shimmering in the breeze, high as my armpits, were the best simulation of a jungle I'd ever seen. Later, when I understood how we planted and mowed alfalfa to create hay for the cattle, I learned to lie down in the fields on hot summer days, letting the plants brush seductively over my body, their chill seep into my flesh.

But that first evening, the tangle of green frightened me. No doubt, my new father mowed the tall alfalfa around the house, since his wife and nine-year-old child were accustomed to short lawns, and feared rattlesnakes. The green hay was so heavy I tripped crossing the windrows. Over my head hundreds of birds dipped and swooped, chasing insects disturbed by the mower. No doubt nighthawks boomed and dived.

In the ranch yard, near the garage and barn, a new house was growing; the old one, raised on jacks, was perched in the nearest hay field on railroad ties for us to occupy during the summer. While my mother was inside unpacking, starting supper, I met my father's pickup coming from his evening tour of the ranch buildings and cattle. We stood outside, enjoying sunset, and I asked about the strange sounds from the sky, something like an explosion of air. He told me about the birds then, but though I can see the scene, I don't really remember it. Other things about that summer in the old house are more vivid.

Since that evening, nighthawks have been my favorite prairie bird. Even then, I seemed to know the Lakota call them "thunderbirds," but how I knew remains a mystery. The metaphor behind the name surprised me, but it became clear during the first autumn thunderstorm I watched with my father.

In August, when ranchers are cutting hay for winter, South Dakota's southwest is baked by the sun most days, bludgeoned by savage thunderstorms, hail, and lightning in late afternoons. My father often repeated stories of neighborhood men killed by lightning on tractors or horses, but he seemed to challenge the storms, to stay out in them longer than necessary for a man so sure of their power to kill.

Later, I wondered if braving the lightning was the only way he allowed himself to shake a fist at divine power, his challenge to the creator to make himself known, a Job-like fatalistic dare. For ranchers and farmers must, above all other lessons, learn patience with unalterable fates: with blizzards during calving season, hail that flattens crops, lightning that kills one expensive bull in a pasture full of elderly cows. Ranch life is crowded with chances for disaster; those who count the awful possibilities, who spend too much time being afraid of what may happen, become paralyzed, unable to act. "Never count the dead ones" became one of my father's mottoes.

He also often said, "What can happen, will." He never screamed curses when machinery broke down, cows refused to go through a gate, or calves brought less money in fall than we'd spent raising them. He faced whatever came, allowing himself only one terror, one chance to act out the anxiety he must have felt. A single wasp buzzing around his head on a summer day could move him out of the slow pace of his days, make him dance, ducking and weaving, waving his handkerchief at the insect, his loose-jointed height suddenly swift and graceful. He admitted his fear of wasps without embarrassment, probably my first lesson that even the brave are allowed fright.

He declined to fear thunderstorms. When I refused to ride a horse under thunderheads from fear of being struck by lightning, he looked contemptuous. If he was haying when a thunderstorm advanced, he wrestled the old tractor and creaking stacker slowly around the field while lightning moved toward us from the Black Hills. I stopped watching falling alfalfa and fixed my eyes on the bubbling underside of the dark clouds overhead. Each time his tractor neared the truck, I reached for my ignition key. Each time, the stacker teeth would drop, scoop up another windrow of hay, and the machine would lumber back toward the stack; I'd shove the throttle ahead and drive faster on the next round, shoulders hunched, expecting a blast between my shoulder blades.

But even then, I noticed the birds, not individuals to me yet, just fluttering shapes that seemed more numerous before a storm. My mower made insects fly up out of the protective vegetation into the winds before the storm, so the birds could snatch them more easily. I felt like a whirlwind, beating insects from hiding to be harvested by birds.

When a lightning bolt plummeted into the hill less than a mile from us, my father would finally pull up beside the stack; I'd lift the sickle bar, shift into fifth, and race toward him. We still had to cover the exhaust stacks of the tractors with tin cans to keep them dry, dash to the truck, drive home, put the truck in the garage, and sprint for the house. By the time we stumbled inside, my mother was terrified, huddled in the dark bedroom as electricity burst over the barn roof. We shouted to be heard over the thunder.

Once inside, father would walk from window to window, watching lightning buffet the hillsides, afraid of a prairie fire. In a year when we'd had ample winter snowfall and spring rain, and were busy harvesting a generous hay crop, he would tell yarns about storms he'd seen as a boy, how the horses ran away with one of his brothers and tipped the mower over, or how an uncle had been killed by lightning. Sometimes the clouds dropped down gray and boiling with water, a smashing flood that scoured the earth instead of soaking into the dry soil. "Well, anyway," he'd say, "it filled the big dam. Tore out the fence, though." Later, as he discussed the storm with other ranchers, they'd all shake their heads in resignation and change the subject.

Watching a windy storm in a dry year, he set his mouth grimly and stared out the windows. "That's a fire-starter," he'd say as a bolt struck, and I'd strain my eyes watching for smoke. He seemed to relax, become more congenial during storms; I think now he was hiding his rage for my benefit. Either lightning would start a fire, or it wouldn't. If it did, either the rain would put it out, or it wouldn't. If the rain didn't put it out, either the volunteer fire fighters would, or they wouldn't, and we'd lose the summer's harvest. Neither anger or fear would change the odds.

Any summer storm could bring hail; during the hours we paced the rooms, we listened rigidly for the first hump on the roof. A truly destructive hail always seemed to begin with one loud whack, followed by a long pause while we guessed the size of that single hailstone. We'd look at one another, and my father would raise his eyebrows and tilt his head, as if the next stone's impact might be inaudible. The second blow was so loud I'd jump, and my mother would scream. The third followed quickly, like a drummer testing his skins before the concert, warming up for the *1812 Overture*. We'd stand silent at the big picture window as jagged chunks of ice battered leaves and branches from the trees, bounced on the driveway, chopped at the base of the tall yard grass we hadn't had time to mow. We never stated the obvious: that any crops we hadn't gathered yet were gone.

At such times, the prairie outside the window looked more like a turbulent ocean; gusts rattled the windows, and gray sheets of water swept across the landscape, so we couldn't see more than a few feet beyond the glass barriers that protected us. Occasionally the murk lifted so we glimpsed one of the huge old cottonwoods east of the house, branches straining with the wind.

Suddenly, above the clothesline, I saw movement, the outline of a bird. I cried out, sure it would be battered to death by the hail. Its wings were beating furiously.

"Oh, that's a nighthawk," my father said, and a moment later I saw the distinctive white spots on its wings. Instead of flying into shelter, or letting the wind take it south, the bird continued to flap in place. Watching, I began to compare the nighthawk's capers with the motions of a human in a hot shower: raising both arms, bending the neck, turning in every direction so the blessed heat can pound muscle aches. The thunderbird seemed to be doing the same, except that it required tremendous energy for the bird to remain in one spot against a fifty-mile-an-hour wind. For a quarter hour it raced furiously in one spot, until it was drenched.

After that, I always saw the nighthawks cavorting in gales, noticed they appeared before the rain to hunt the wind-beaten grass.

My first intimate look at a nighthawk came years later, when I saw one tidily balanced on a gate post. Eyes closed, it didn't move while George drove the pickup through the gate, and I pulled it shut. Then we tiptoed close. "Is it dead?" he asked.

When our noses were four inches from its beak, one eye opened, then slowly shut. We memorized details of its sleek body, longer than George's open hand, about nine inches from beak to tail. Dappled gray and tan feathers made it nearly invisible against the weathered wood. Sitting still, its posture concealed white patched under its chin, and on each wing. In flight, the white spot in the middle of each wing identifies the bird's family; no other prairie flyer displays them.

Once I'd seen the bird's storm performance, I always called him the thunderbird, nearly forgetting that most folks know him as a nighthawk, and a specialized few as *Chordeíles minor*. Technically, the bird is not a hawk but a goatsucker, from a family certainly named for habits other than those shown on the plains; we're not fond of goats in cow country, though we refer to pronghorn as "goats." But even the speedy nighthawk would have a hard time sucking antelope.

Like his cousins the whippoorwill, poorwill, and chuck-wills-widow, the nighthawk has an enormous mouth, and specializes in catching insects on the wing, a practice called "hawking." Birds of prey like true hawks kill larger mammals by shock and speed; the nighthawk, really not a hawk at all, performs the same skillful operation with tiny, zipping insects.

Several times over the years, riding a horse or walking, I've found a nighthawk's home; no one could sensibly call it a nest. The bird simply lays two eggs blotched with shades of green on an outcropping of limestone. The rough stone camouflages the eggs, and perhaps holds warmth while the parents hunt.

Sometimes, when I walk among my windbreak trees and bushes, a nighthawk flutters up from the ground just ahead of me, and I glimpse one mottled egg beside a down chick, both yellow splotched with green. Invariably, an adult drifting just above the tallest grass tricks my eyes away, and when I look again for the chick, it is invisible. Yet in two more steps, my foot might crush it. Once I know the locale of a nest, I walk wide around it for a month, until the evening when I notice young, clumsy nighthawks bumbling after an adult, trying to gulp insects.

After the nighthawk's flight became familiar, I could spot them at a glance. At first, they seem to fly like other birds, with even, rhythmic wing strokes against the air. Then I noticed that hunting nighthawks alter the tempo; several very brisk strokes push their bodies up, and alternate with strokes at normal speed. They mount the sky in jerky motions, uttering a single high note, until they are barely visible. When I spot the outline of a nighthawk ascending, I watch for what follows. From overhead, so high it may be invisible, the nighthawk drops nearly to the ground in utter silence. At the end of the dive it swoops up, and a peculiar boom reverberates.

How, I wondered, do they make that eerie tone? Does it come from their throats, or is it produced by the intricate wing motions demanded to curb that furious dive, and soar again? One guidebook said the wings produce "a peculiar musical hum" in the courtship dive. But, though the sound occurs during courting season, I also heard it on summer nights as dozens of the birds hunted around my eaves. Later, when the adults sported with adolescents, the sounds multiplied; certainly they were not courting so late in the season. Several source books fail to mention the sound at all, and yet it's the surest way to identify the flying acrobat of the prairie, even if they're hidden by darkness. No other creature on the prairie makes a sound that can be confused with it.

During the long autumn twilight evenings in the month following my husband's funeral, I often sat by the stone cairn he'd built on the hillside, watching the sun suck daylight beyond the western hills. Almost hidden in the tall grass, I became invisible, or at least insignificant, to nighthawks swooping after millers, mosquitoes, and grasshoppers. As darkness gathered, it seemed to surround me with the breath of nighthawk sounds, like great horns blown on a high mountain. Christians imagine the awe-inspiring tones of heavenly trumpets, but I was thrilled by knowing that scrape of night on crescent wings made the night hum.

That fall, the millers were particularly abundant, infesting every corner of the house. Often, as I lay in bed, a miller squirmed over my face, wings trembling against my cheeks; I'd wake clutching my face, sneezing and muttering, my revulsion doubled by knowing each living miller meant more cutworms chewing garden plants and flowers. When my loneliness for George grew acute, I invented desperate entertainment. Each sultry evening I turned on the water hydrant and directed a stream of water at the outside walls of the house to drive the millers out from under the siding. When millions of them fluttered around the house, befuddled, lurching, and spinning on wet wings, the sky would fill with nighthawks wheeling, swooping, snatching. Often a nighthawk would light on the porch railing with three or four millers in its mouth, nearly overbalanced by its catch. As amusement, it was trifling, and perhaps morbid, but I'd harass the millers until dark, hypnotized by the aeronautics of the birds.

When my hands turned stiff and aching from the cold water, I'd curl up in the hammock to sit motionless until the house was a black square against the deep night. Gradually my eyes widened, gathering light residue until I could see clearly. Often a nighthawk's sleek shape would glide out of dusk to stare at me from the porch railing. Frustrated by seeing only a silhouette, I would remain motionless as long as it stayed. The whoosh and boom of nighthawks dropping out of the stars to catch insects echoed everywhere.

Later, after midnight emptied the house of life and sound, turned every painting and shape to the face of death, and detached me from mankind outside my walls, I'd go in my nightgown to the deck or hillside. I'd sit with Frodo nudged close, his ears swiveling; we'd listen to nighthawks rushing through the dark, killdeer calling by the water, coyotes warbling, assuring ourselves that we were not alone.

Life on the ranch, even after George's death, was charged with such experiences, even though my father and I worked harder than ever. At night, reliving them, I recorded them in my journals, storing up memories, material that might someday provide essays and poems. Readers of my year's journal, *Windbreak*, had told me they loved the book because I wrote about ranching, a subject unfamiliar to many in the urbanized world. Some wanted me to publish a journal of another year. Everyone agreed that only by staying on the ranch would I continue to write.

Yet one spring day I woke in a small house in a metropolitan suburb, at the end of a winding trail of actions and reactions, at the beginning of the unknown. Lying in a strange room, I recited facts. I'd spent the past forty years learning how to work hard on a ranch; absorbed its lessons, discovered myself pledged soul, body, and brain to a few thousand acres of buffalo grass on the arid high plains, to a ranch.

Eyes open in that room, I stared at the reality that I was nearly fifty, widowed almost five years. My first book had been in print eight years; the fifth, seven months; I was negotiating to hire an agent for the next. All of my writing concerned ranching. My father, eighty-three, memory and temper disintegrating, wanted me where he could summon me, night or day, to help him. If I didn't answer the telephone on the second ring, he hung up and drove to my house. He refused to read any of my writing, to acknowledge that I might have other obligations and pursuits. But he would not formalize my relationship; I was his daughter, but not his heir, his partner, or his manager. If I trusted him, he said, we needed no contract.

For forty years I had been learning by working beside him; eager to prove I was "as good as a man," I'd lifted beyond my strength, and old injuries made heavy work painful and sometimes dangerous. He acknowledged that the ranch work was getting beyond the abilities of "an old man and a crippled girl," and found ways to keep me from doing the work I loved most, riding horseback. But he would not hire help, or listen to my suggestions for change, or acknowledge that I knew enough to run the ranch. My jobs were those I'd done at the age of ten. Some days he insisted that he had supported me all my life.

Life on the ranch became a turbulent struggle to help him in spite of himself, while his decreasing abilities began to destroy the ranch he had built. When I was gone, he couldn't remember where or why; when I was home, he refused to ask my help, insisting that I should be with him all day in case he needed me. I was frustrated and angry at my inability to right the situation.

When I woke in that suburban bedroom, to the wakeup call of screeching tires, railroad whistles, and choking smog, I had no strategy. Living by my wits, I bargained for time to reflect, find resolution for my dilemmas. Forty years had passed since I'd lived in a city, not counting college and graduate school; I knew those residences were temporary. The room where I woke that spring morning stood one-quarter mile downwind from an oil refinery; not long after I moved my toothbrush, an explosion sent burning debris a half-mile in the air. Blazing day and night, a fire designed to incinerate unused vapors flavored every breath with crude oil, and reflected in the bedroom windows and mirrors at night to wake me from nightmares.

A block away, a school gushed screaming elementary children each afternoon; a few minutes after arriving home, most of the tots emerged with remote control cars, racing them on the sidewalk and street until dark. Three blocks away rumbled a railroad switchyard. Across the street lived several young men who broadcast their collection of rock and roll tapes night and day, next door to a trucker who worked at night. During the day he mowed his lawn, and repaired engines. Across the street lived a man with a pickup truck wearing a bumper sticker proclaiming it saved by Jesus; the engine told a different story. On weekends, he peered under the hood while the engine ran full throttle; perhaps he prayed for its soul.

Each house in the neighborhood was equipped with a large dog, penned in the back yard; each canine protested his incarceration loudly each time a car or person passed, and each bark was a signal to every other dog for blocks in all directions. Most persistent, most energetic, and most annoying, were the puny yapping squeals of tiny dogs with no apparent need to pause for breath.

I tucked my computer into a spare bedroom, piled books on the floor and bed, kept files in the closet, and stuffed foam plugs in my ears to write or to sleep. Instead of walking on my hillside collecting wildflower seed, watching nighthawks, ducks, coyotes, and cattle, and watering trees, I bicycled or walked on city streets, and trained Frodo to walk with a leash. A test of my powers of adaptation, I told myself; many artists have created brilliantly with less.

Friends responded to news of my migration with shock and horror; some of those who share my view that our natural environment is worth protecting recoiled as if I were wearing a Dow Chemical T-shirt and drawing a paycheck stained with oil, or blood. Some told me stories of writers who produced a book or two, and then disappeared from the public eye into alcoholic obscurity. In the city, I searched for fragments of prairie to make me feel at home. During the day, blackbirds patrolled back yards, hunting food. Each evening, they congregated to squawk and shrill in a tree behind the rock music fans' house, perhaps drawn by the competition. One morning two landed on the electric wire over our small back yard; they fluttered and chattered for a moment or two before the female flew down to scratch among the bachelor buttons, Sweet William, and other wildflowers I'd transplanted from my prairie hillside. The male remained overhead, watching alertly, grooming his wings. Shortly, another blackbird lit a few feet away on the overhead wire. Instantly the male flew into the air, screeching, and dived toward the intruder, who promptly flew away. The female, apparently positive of her protection, kept digging. Such confidence in the male protector; I envied her.

One day Frodo's barking drew me to the back door; he stood at the fence, shouting alarm at several neighborhood children walking by. I tried to explain that the alley was not his territory, but he's used to guarding two ranch houses a quarter mile apart, a half mile of road, and several miles of pasture fence; he takes his duties seriously. A few minutes later, his indignant yaps interrupted me again; this time a robin was bathing in his water dish. After thoroughly soaking itself, the bird hopped up on the rock I'd put in the water as an escape route for small animals, shook water from his feathers, and groomed himself, ignoring the dog.

Once, I saw a kestrel cruise past, and thanked him for appearing in my life. Each night, sitting in a chair in the back yard, or digging space for more flowers, or watering the lawn, I listened half-unconsciously for the boom of nighthawks. Each night, watering and weeding the wildflowers growing from prairie seed, I'd scan the sunset sky above the refinery, hoping to see the familiar crescent shape, the peculiar way a nighthawk's wings flutter as he mounts high before his striking fall for prey. Each night I watched the blackbirds gather, noticed robins and sparrows; once I even saw a meadowlark in a field a few blocks away. These tiny signs somehow raised my spirits, gave me hope.

On a hot afternoon in August, we were sitting in the back yard, visiting with friends. We'd proudly shown them the six tomato plants, heavy with fruit, and pulled rhubarb for them to take home; I was laughing at myself for preserving these tokens of prairie hospitality even in suburbia, but I was proud, too, at how much country I'd managed to tuck into a tiny back yard.

During the day I had written a draft of this essay, but I was weary, unsatisfied; I longed for the prairie hills around my home, wondered how the cattle looked. Each week, I talked to my mother and father on the telephone; every two weeks I went home to help my father move cattle, to brand, to handle jobs that he could not do alone. My mother repeatedly urged the only solution she could see, "Why don't you just come home? You father's always right."

When I heard the familiar *peenk* of a nighthawk, I looked up automatically. Overhead wheeled dozens, spinning and dipping, whirling and fluttering. Breathless, I realized the house seemed to be the center of a flock; the rest of the sky was empty. With tears in my eyes, I watched the familiar shapes swoop, and heard their booming rush after prey.

Our friends were puzzled; crying over birds? I tried to explain how odd it was to see nighthawks here, and walked to the alley fence to stand among the relocated wildflowers staring, hiding my tears, engraving the sight on my mind. An omen, I decided; perhaps my confusion would end, some solution would be found, life would go on.

That night we attended a concert by Mason Williams' bluegrass band. As the fiddle strokes of "Cotton-Eyed Joe" pulsed through the hall, I closed my eyes. The high, sweet notes spiraled and glided beneath the domed ceiling like the nighthawks' flight. The music seemed to resonate in my bones, vibrate in each nerve and muscle. I shivered, and looked at my companion's watch; 10 P.M.

In that instant, in the kitchen of the ranch house three hundred miles away, my father collapsed and died.

Over a single house in a large city, hundreds of nighthawks flew high and sweet as fiddle music.

—9—
Apologia

Barry Lopez

A few miles east of home in the Cascades I slow down and pull over for two raccoons, sprawled still as stones in the road. I carry them to the side and lay them in sun-shot, windblown grass in the barrow pit. In eastern Oregon, along U.S. 20, black-tailed jackrabbits lie like welts of sod—three, four, then a fifth. By the bridge over Jordan Creek, just shy of the Idaho border, in the drainage of the Owyhee River, a crumpled adolescent porcupine leers up almost maniacally over its blood-flecked teeth. I carry each one away from the tarmac into a cover of grass or brush out of decency, I think. And worry. Who are these animals, their lights gone out? What journeys have fallen apart here?

I do not stop to remove each dark blister from the road. I wince before the recently dead, feel my lips tighten, see something else, a fence post, in the spontaneous aversion of my eyes, and pull over. I imagine white silk threads of life still vibrating inside them, even if the body's husk is stretched out for yards, stuck like oiled muslin to the road. The energy that held them erect leaves like a bullet; but the memory of that energy fades slowly from the wrinkled cornea, the bloodless fur.

The raccoons and, later, a red fox carry like sacks of wet gravel and sand. Each animal is like a solitary child's shoe in the road.

Once a man asked, Why do you bother? You never know, I said. The ones you give some semblance of burial, to whom you offer an apology, may have been like seers in a parallel culture. It is an act of respect, a technique of awareness.

In Idaho I hit a young sage sparrow—*thwack* against the right fender in the very split second I see it. Its companion rises a foot higher from the same spot, slow as smoke, and sails off clean into the desert. I rest the walloped bird in my left hand, my right thumb pressed to its chest. I feel for the wail of the heart. Its eyes glisten like rain on crystal. Nothing but warmth. I shut the tiny eyelids and lay it beside a clump of bunchgrass. Beyond a barbed-wire fence the overgrazed range is littered with cow flops. The road curves away to the south. I nod before I go, a ridiculous gesture, out of simple grief.

I pass four spotted skunks. The swirling air is acrid with the rupture of each life.

Darkness rises in the valleys of Idaho. East of Grand View, south of the Snake River, nighthawks swoop the road for gnats, silent on the wing as owls. On a descending curve I see two of them lying soft as clouds in the road. I turn around and come back. The sudden slowing down and my K-turn at the bottom of the hill draw the attention of a man who steps away from a tractor, a dozen yards from where the birds lie. I can tell by his step, the suspicious tilt of his head, that he is wary, vaguely proprietary. Offended, or irritated, he may throw the birds back into the road when I leave. So I wait, subdued like a penitent, a body in each hand.

He speaks first, a low voice, a deep murmur weighted with awe. He has been watching these flocks feeding just above the road for several evenings. He calls them whippoorwills. He gestures for a carcass. How odd, yes, the way they concentrate their hunting right on the road, I say. He runs a finger down the smooth arc of the belly and remarks on the small whiskered bill. He pulls one long wing out straight, but not roughly. He marvels. He glances at my car, baffled by this out-of-state courtesy. Two dozen nighthawks career past, back and forth at arm's length, feeding at our height and lower. He asks if I would mind—as though I owned it—if he took the bird up to the house to show his wife. "She's never seen anything like this." He's fascinated. "Not close."

I trust, later, he will put it in the fields, not throw the body in the trash, a whirligig.

North of Pinedale in western Wyoming on U.S. 189, below the Gros Ventre Range, I see a big doe from a great distance, the low rays of first light gleaming in her tawny reddish hair. She rests askew, like a crushed tree. I drag her to the shoulder, then down a long slope by the petals of her ears. A gunnysack of plaster mud, ears cold as rain gutters. All of her doesn't come. I climb back up for the missing leg. The stain of her is darker than the black asphalt. The stains go north and off to the south as far as I can see.

On an afternoon trafficless, quiet as a cloister, headed across South Pass in the Wind River Range, I swerve violently but hit an animal, and then try to wrestle the gravel-spewing skid in a straight line along the lip of an embankment. I know even as I struggle for control the irony of this: I could pitch off here to my own death, easily. The bird is dead somewhere in the road behind me. Only a few seconds and I am safely back on the road, nauseous, light-headed.

It is hard to distinguish among younger gulls. I turn this one around slowly in my hands. It could be a western gull, a mew gull, a California gull. I do not remember well enough the bill markings, the color of the legs. I have no doubt about the vertebrae shattered beneath the seamless white of its ropy neck.

East of Lusk, Wyoming, in Nebraska, I stop for a badger. I squat on the macadam to admire the long claws, the perfect set of its teeth in the broken jaw, the ramulose shading of its fur—how it differs slightly, as does every badger's from the drawings and pictures in the field guides. A car drifts toward us over the prairie, coming on in the other lane, a white 1962 Chevrolet station wagon. The driver slows to pass. In the bright sunlight I can't see his face, only an arm and the gesture of his thick left hand. It opens in a kind of shrug, hangs briefly in limp sadness, then extends itself in supplication. Gone past, it curls into itself against the car door and is still.

Farther on in western Nebraska I pick up the small bodies of mice and birds. While I wait to retrieve these creatures I do not meet the eyes of passing drivers. Whoever they are, I feel anger toward them, in spite of the sparrow and the gull I myself have killed. We treat the attrition of lives on the road like the attrition of lives in war: horrifying, unavoidable, justified. Accepting the slaughter leaves people momentarily fractious, embarrassed. South of Broken Bow, at dawn, I cannot avoid an immature barn swallow. It hangs by its head, motionless in the slats of the grill.

I stop for a rabbit on Nebraska 806 and find, only a few feet away, a garter snake. What else have I missed, too small, too narrow? What has gone under or past me while I stared at mountains, hay meadows, fencerows, the beryl surface of rivers? In Wyoming I could not help but see pronghorn antelope swollen big as barrels by the side of the road, their legs splayed rigidly aloft. For animals that large people will stop. But how many have this habit of clearing the road of smaller creatures, people who would remove the ones I miss? I do not imagine I am alone. As much sorrow as the man's hand conveyed in Nebraska, it meant gratitude too for burying the dead.

Still, I do not wish to meet anyone's eyes.

In southwestern Iowa, outside Clarinda, I haul a deer into high grass out of sight of the road and begin to examine it. It is still whole, but the destruction is breathtaking. The skull, I soon discover, is fractured in four places; the jaw, hanging by shreds of mandibular muscle, is broken at the symphysis, beneath the incisors. The pelvis is crushed, the left hind leg unsocketed. All but two ribs are dislocated along the vertebral column, which is complexly fractured. The intestines have been driven forward into the chest.

The heart and lungs have ruptured the chest wall at the base of the neck. The signature of a tractor-trailer truck: 78,000 pounds at 65 MPH.

In front of a motel room in Ottumwa I fingerscrape the dry stiff carcasses of bumblebees, wasps, and butterflies from the grill and headlight mountings, and I scrub with a wet cloth to soften and wipe away the nap of crumbles, the insects, the aerial plankton of spiders and mites. I am uneasy carrying so many of the dead. The carnage is so obvious.

In Illinois, west of Kankakee, two raccoons as young as the ones in Oregon. In Indiana another raccoon, a gray squirrel. When I make the left turn into the driveway at the house of a friend outside South Bend, it is evening, hot and muggy. I can hear cicadas in a lone elm. I'm glad to be here.

From the driveway entrance I look back down Indiana 23, toward Indiana 8, remembering the farm roads of Illinois and Iowa. I remember how beautiful it was in the limpid air to drive Nebraska 2 through the Sand Hills, to see how far at dusk the land was etched east and west of Wyoming 28. I remember the imposition of the Wind River Mountains in a hard, blue sky beneath white ranks of buttonhook clouds, windy hayfields on the Snake River Plain, the welcome of Russian olive trees and willows in creek bottoms. The transformation of the heart such beauty engenders is not enough tonight to let me shed the heavier memory, a catalogue too morbid to write out, too vivid to ignore.

I stand in the driveway now, listening to the cicadas whirring in the dark tree. My hands grip the sill of the open window at the driver's side, and I lean down as if to speak to someone still sitting there. The weight I wish to fall I cannot fathom, a sorrow over the world's dark hunger.

A light comes on over the porch. I hear a dead bolt thrown, the shiver of a door pulled free. The words of atonement I pronounce are too inept to offer me release. Or forgiveness. My friend is floating across the tree-shadowed lawn. What is to be done with the desire for exculpation?

"Later than we thought you'd be," he says.

I do not want the lavabo. I wish to make amends.

"I made more stops than I thought I would," I answer.

"Well, bring this in. And whatever I can take," he says.

I anticipate, in the powerful antidote of our conversation, the reassurance of a human enterprise, the forgiving embrace of the rational. It waits within, beyond the slow tail-wagging of two dogs standing at the screen door.

—10—
On the Daurian Steppe

A Mongolian Journey in Search
of the White-Naped Crane

Peter Matthiessen

Northwest from Beijing, China's teeming plain is left behind, the soft farm greens giving way to darker forest greens of sharp small mountains where the Great Wall winds like a serpent across ridges. Soon the mountains descend to the drier, less domesticated landscapes of Inner Mongolia, which subside in turn into the harsh grays and yellows of the Gobi Desert. Within the first hour of the flight from the ancient Chinese capital, one is well into the Central Asian wastes of southeastern Mongolia.

Gradually, the desert is displaced by semi-desert plains, then grassy steppe, crossed here and there by veins of darker green where seasonal streams make their way over the plateaus. Here and there as the plane descends, grazing animals tended by lone riders cross the landscape—horses, cattle, sheep, and goats, a camel; and loose clusters of herdsmen's round felt tents or yurts, called *gers*, crop up, as white as mushrooms. The herds and riders, and *gers*, too, drift to the very edges of the glinting city. This is Mongolia's capital, Ulan Bator, which lies on the Tuul River in the north of the great steppe, more than 4,000 feet above sea level.

Mongolia's culture, originally a wandering shamanistic one in the thrall of local deities and the sky god Tengri, has been mainly Buddhist since the late sixteenth century, when the Gelugpa sect of Tibetan Buddhism (the sect headed by the Dalai Lama) became the state religion. Even then, the society remained nomadic, independent, with a warrior spirit. Quite possibly, Buddhist influences softened this spirit, for the Mongol herdsmen were eventually absorbed by the Chinese empire. Not until the last emperor was retired, in 1911, did northern, or "Outer," Mongolia reclaim its independence. This meant, in effect, turning to Russia for protection, and in 1921, the new Union of Soviet Socialist Republics used the founding of a "Mongolian People's Party" as a pretext to invade the country and set up a satellite Communist government. For seven decades, Mongolia remained a puppet of the USSR.

In 1989, with the departure of most of the Soviet soldiers, came official restoration of both Buddhism and Genghis Khan as "the priceless cultural heritage of the Mongolian people," and in September 1991 the Dalai Lama arrived on the wings of a religious yearning so pent up by decades of oppression that tens of thousands came to welcome him, arriving by bus and truck and horseback, raising spiritual dust all over Mongolia. Most Mongolians are still nomadic, and the half million people in Ulan Bator— almost half of them camped behind wood fences, in *gers* that can be taken down and moved back out onto the steppe in about an hour—represent a quarter of the entire population of a vast country the size of Western Europe. Thus much of Mongolia remains all but empty, not only the deserts to the south but the remote marshes in the far northeast, where endless horizons of Daurian Steppe stretch away over the frontiers with Inner Mongolia and Siberia.

With the collapse of the Cold War, regions of Asia have been opened to visitors from the West for the first time in many years. Among these visitors have been ornithologists from all over the world who are, rightly, concerned about the cranes of Asia. The crane family first appeared on the fossil record some sixty million years ago, in the time of dinosaurs, and modern cranes have flown the earth for perhaps nine million years—the oldest (and tallest) flying birds still in existence. Among the fifteen crane species, nine are breeding birds of Asia, and of these nine, at least five are already considered rare and appear on most official lists as "threatened" or "endangered," due largely to degradation and destruction of their wetland habitat. (They are also diminished due to hunting by various predators, man in particular, and to the disruption of both breeding and winter territories by man's ever-proliferating activities, such as war.)

These five uncommon cranes include the extraordinarily beautiful white-naped crane, *Grus vipio*, which migrates south from its known breeding range in the Amur River region of northern China and Siberia to its main winter sites in southeast China, Korea, and Kyushu, the great southern island of Japan. Before any species can be systematically protected, its breeding grounds and migration routes must be well monitored, and the white-naped crane presented a special problem, since it has long appeared on its winter grounds in markedly greater numbers (its roughly estimated total population is about 4,500) than its known breeding grounds could possibly account for.

Because in the Cold War decades the isolated Asian ornithologists had little opportunity for fieldwork, or travel to other countries, or even good communication with colleagues in the West, the list of known ranges of the far-flung cranes remained woefully incomplete. Then, in 1991, with the Soviets gone from Mongolia, the International Crane Foundation (ICF) in Baraboo, Wisconsin, received an exciting report from two Mongolian ornithologists that *Grus vipio* was breeding in remote regions of the Daurian Steppe and that others had been sighted along the wild Kerulen River to the south, near the Chinese border, although no Kerulen nesting site had been discovered. Last summer, the Kerulen River was the first destination of the ICF's international expedition to Mongolia, organized to verify and explore any new breeding grounds of the white-naped, or Daurian, crane. It was this expedition that had brought me to Mongolia.

On the afternoon of July 21, James Harris, deputy director of the ICF, Tsuyoshi Fujita, a young Japanese ornithologist of the Wild Bird Society of Japan, and I were met at the Ulan Bator airport by the organizer of the expedition, the Mongolian mammalogist S. Chuluunbataar, a man of sad demeanor and soft dejected voice that hide a sudden delighted smile and a sense of humor. Chuka, as he is called by friends, was accompanied by the two ornithologists—Drs. Ayurzaryn Bold and Natsagdorjin Tseveenmyadag—who had first reported to the ICF that *Grus vipio* was a breeding bird in northeastern Mongolia. Chuka speaks English, and "Go" Fujita has some English, too, and Go and the Mongolians share some scraps of Russian. After years of enforced sociability in Asian lands whose languages he does not speak, Jim Harris has perfected a hearty missionary manner that counts on an unflagging smile and voluminous laugh. But the good nature and enthusiasm are sincere, and behind them lie sound ornithological knowledge. So there was to be communication of a kind. Otherwise, we were to get by on goodwill, Latin nomenclature, and an intense shared passion for wild cranes—which could have served as a metaphor for all that is wild and beautiful and waning in the realm of nature.

Mongolia in summer is notorious for hard and sudden changes in the weather, and late that first afternoon we left Ulan Bator in a storm of rain and hail that brought mud rivers in flash floods through the streets, stranding cars and stalling buses and causing the goats to press against the walls, yet in no way fazing the pedestrians, who scarcely quickened their pace in search of shelter from this cold deluge.

We traveled in two four-wheel-drive vehicles with spirited and obliging drivers, Dorj and Dawadav, who assisted with the cooking and the camp. One of the vehicles dragged a two-wheel trailer containing tents, a stove, food, tools, and the like. Originally, the expedition was scheduled to fly east from Ulan Bator to the town of Choybalsan, near the border with China, then travel overland down the Kerulen, but due to fuel shortages, all domestic flights had been canceled until further notice. A four-day drive was made instead on what one reference book refers to as "the east-west highway," which turns to mud and ruts a few miles east of Ulan Bator. "We hope you will be tough enough to make this journey," said the taciturn Dr. Bold, who, at fifty-six, is Mongolia's pre-eminent ornithologist, a rough old badger in dark glasses and brown business suit that he continued to wear (without the tie) on our safari. His thirty-seven-year-old associate, Dr. Tseveenmyadag, called Tseveen, is just as competent and hardy, as he would demonstrate unobtrusively throughout the journey. Perhaps because he is shy and quiet, the retiring Tseveen wore a camouflage suit, but the suit's unusual tropical greens made him a bold feature of the grassland landscape.

An hour from the capital the dirt track climbed a high valley to the continental divide. At the pass, a large cairn, or *obo*, stood adorned with rags, liquor bottles, rusted cans, broken thermoses, and other miscellaneous offerings, including two pairs of old crutches. There were also a few tattered Buddhist prayer flags. Though traditional mountain cairns set up by wayfarers to placate intemperate deities predated Buddhism, this one appeared to be a debased form of the Buddhist stupas or chortens found on passes throughout the Himalayas. Asked if this disreputable *obo* represented the disdain for Buddhist ways beaten into the Mongolians by their Russian mentors during the past half century, Chuka said no, and Bold agreed. It was only more evidence, they said sadly, of very low morale among the people.

Long after dark, we pitched our tents in a small grassy basin. Here on the steppe, the July night was cold. Scorpio and Cygnus bristled with light and the Milky Way descended to the horizon. In the morning, the first Mongolian lark I ever saw awaited me on a grass tuft just before the tent.

I found puffballs and some mushrooms that Dorj cooked up for breakfast. Nearby steppe marmots stood upright at their burrows like huge prairie dogs, and along the track a sick one lay, gold and ivory fur rising and falling with its breath in the morning sun. A steppe fox whisked through the grass, and a little farther on, a steppe eagle swooped on a young marmot. Unwilling to relinquish life, the eagle glared at us as it tore free red shining shreds of meat. Neatly, then, it eviscerated the creature, leaving the heavy guts behind as it dragged the rest away over the grass. In the high tablelands farther east, a pair of elegant gray birds traversed a barley field, hurrying one chick. These were the demoiselle cranes that I first saw many years ago on winter range in the Sudan. Unlike cranes of the genus *Grus*, such as the white-naped, the demoiselle crane, *Anthropoides virgo*, lacks the crown of carmine skin that makes those larger birds so striking, but the very simplicity of its markings—silver gray and black—set off by a silken bib of black plumes at the breast, make it one of the most beautiful of its family. (The demoiselle received that name from Queen Marie Antoinette, who admired the delicacy of her captive specimens.)

At a signal from the parents, the demoiselle chick pressed itself flat in the thin barley. The adults moved away, affecting to feed, then took off with a ratcheting horn-note call. Almost at once they circled back, alighting at a little distance, "craning" in distress in our direction. For we had found the flattened chick, like a gray pullet with elongated legs and enormous feet. Go banded the chick—the first crane that we banded on the expedition— and it ran off wearing a green leg band, Y-01.

In wildflower prairie, the track followed a small stream, arriving eventually at a lake and mineral spring where fifteen or twenty demoiselles stalked the surrounding grassland. Though nights were cold, broad day was hot, and in the spring a few families of nomad herdsmen were smearing their bodies with medicinal mud under the patient gaze of waiting horses. Already the plateaus were subsiding into the high rolling plain of the open steppe, islanded by white dots of solitary *gers* with their dark strings of horses. Soon the stream crossed broad wet meadows of river floodland. Here the Kerulen River, not deep but cold and very swift with the quick loss of altitude, made a great bend toward the east.

All afternoon the demoiselles increased in number; in one flock we counted sixty or more, and captured and banded a pair of chicks, as six Tatar boys came galloping up to watch the operation. Some were riding bareback in their well-soiled cloaks and dusty boots, their skin sun-blackened and hair uncropped. Fighting back their own awe and astonishment, they watched Go closely as he opened his shamanic kit of new and glittering equipment, drew a white muslin bag over one bird's head, taped its legs and jacketed it to keep it still, fit and glued broad green plastic rings to an upper leg, then measured and weighed the bird while the glue dried. With unrelenting precision Go inscribed the last detail of his data in his book before cutting the tape removing the bag and jacket, and releasing the chick to run away over the plain.

Asked by Chuka if they caught young cranes, and if they ate them, the boys said proudly, Yes, we do!, but perhaps this was bravado. Any chicks would be easily caught by boys on horseback, yet the demoiselles seemed quite unwary, drawing near man's habitations and his herds with a confidence suggesting that they were rarely hunted.

We camped that night on the riverbank, to the calls of cranes and Eurasian cuckoos, red-shanks and an owl, replaced at dawn by the alarmed bark of marmots and the skylark's daybreak song. The day began with fitful rain and cold. Since this was the season of monsoon—Mongolia gets almost all of its precipitation in midsummer—we anticipated more rain and rather dreaded it. Already the red ruts of the east-west highway were so rough that earlier travelers—mostly trucks, to judge from the very few vehicles we saw—had forged their own new tracks to avoid the holes, and in places, as many as ten or a dozen routes fanned out, mile after mile, before braiding once again where the highway forded a stream or crossed a hillside.

This is a landscape of great saker falcons—two adults flying, then two juveniles on marmot mounds, then a single bird, another, and another, all within the space of a few miles. The big sandy bird has the heavy flight of a gyrfalcon—the redoubtable peregrine would appear "light" by comparison. In taking to flight, the saker's sharp wings cut the air like two knives sharpened one against the other.

Though most of the steppe is gently rolling, with plateaus, there are glimmering flat regions where the horizon seems so near that giant horsemen rise between the grass and sky. In their traditional Mongol hats and cloaks, called *deels*, the horsemen appear mythic, and the horses seem more ancient still, with their chunky bodies and thick legs and the crude lumpy heads of *Eohippus*. Some of their goats, too, look archaic, with crazy yellow eyes and outsize horns, whereas the sheep and cattle they herd appear modern—that is to say, dulled and stupid in appearance and comportment, befitting animals that have given up all sense of anything but satiation, content to be shunted about in bawling herds. The two-humped camels are used almost entirely as draft animals, dragging small two-wheeled covered carts jammed with household possessions when the grazing gives out and the time comes to move the herds and *gers*. Once a new camp is made, these carts, tilted forward onto their shafts and parked in a line outside the *ger*, are used as storage bins and hitching posts for one or more horses, which, in this horse culture, are kept saddled from daybreak until nightfall. Since but one percent of the land in Mongolia is arable, and since both rainfall and the date of the last frost in spring are as capricious as this summer weather, field agriculture is of minor importance.

Later that day, less than an hour east of the settlement of Ondörhaan, the axle sheared and one wheel spun off the two-wheel wagon, bouncing and leaping behind Dorj's car, bringing our expedition to a sudden halt. With Dorj and Tseveen, Chuka set off for a mining settlement fifteen miles away, while Bold guarded the dumped cargo and Harris, Go, Dawadav, and I took advantage of the delay to run down, catch, and band two chicks of some unsuspecting demoiselles that had drawn too near the scene of our calamity.

On a nearby hilltop stood a *ger* with a large corral, and inevitably its master, a big Mongol in a night-blue *deel* with orange sash, mounted his horse and with his boy and his ancient hammer rifle rode down to inspect the cart and commiserate over the vagaries of fate. Having then remounted and ridden away, crossing the hard plain at the stiff-legged run, not quite a gallop, which tough Mongol horses can maintain for hours, the pair returned toward twilight with a fresh-killed marmot, a local delicacy, which they flung down before our camp as a gracious token of steppe hospitality. Not until long after dark did the others return with a tough old cart bought at the mining village, and so we made camp that night near the track, knowing that out here on the east-west highway no vehicle would disturb our sleep.

Following a breakfast of fatty marmot, served cold with rice and tepid coffee, we followed the track east through grass that, though still utterly treeless, without so much as a poor shrub, was becoming taller and less sparse, with a gradual change in the bird fauna. The cranes were fewer in this long-grass country, and there were no falcons. Only the wheatears and the larks seemed undiminished in their numbers.

That night we camped east of Choybalsan, the only town of any size in the eastern part of Mongolia, and the next day headed north and east across the plateaus above the Kerulen river plain, over empty grassland without *gers* or horses. Birds of all kinds seemed very scarce, and a solitary pair of demoiselles, without a chick, were the only cranes we would record. However, we did see a crane relative called the great bustard, which carries itself like a short-necked, short-bodied, and short-legged crane—less graceful, perhaps, but just as dignified. Flaring white on the wings, the bustard lifted heavily from the thick grass and pounded away over the steppe. This huge steppe bird, once relatively common to Russia but now almost gone, is one of the rare species we had hoped to find in these wilder regions of Mongolia.

Where thick grass eventually gave way to short sparse growth, there appeared a flock of more than sixty demoiselles. At this season, the flocks are mostly composed of immature birds as well as pairs that failed to breed or bred but lost their chicks to raptors or mammal predators. Perhaps these cranes were already assembling in the first stirrings of the migratory instinct that will lift them up one September day and beat them south on a long journey across the central Asian deserts, the Altai Range, and the great Takla Makan Desert, the remote Kunlun Mountains and Tibet, then across the Himalayas to the Ganges plain. Many will soar down the deep gorge of the Kali Gandake River, which parts the snowy peaks between Annapurna and Dhaulagiri, and some will be carried high by its great winds. The elegant and airy demoiselle has been recorded five miles above sea level, passing the shining peak of Annapurna.

The track bore east over the plateau, and below, great curves of the Kerulen appeared. This river plain that was our first destination looked wild and all but uninhabited, yet appeared too dry to attract the white-naped crane. But as the track moved close to the reedy margins, the plain came alive with birds—swan geese and greylags, spot-billed ducks, herons and cormorants, terns and gulls, myriad sandpipers. On a grassy site overlooking a broad oxbow lagoon known as Tsagaan Sum (White Church Lake) stood a strange mound of earth and rubble—all that remains of a Buddhist temple torn down in this place during the great anti-clerical purges of the 1930s. The steppe tracks and this mute heap, circled by swallows, were the only signs of man in the broad landscape.

On a high bank along a deep bend in the river, Bold and Tseveen located their old camp of a few years ago, when they had recorded their first sighting of the white-naped along the Kerulen. Bold produced a set of small red cups and a bottle of arak, as we listened to the murmur of the river, and the wail of lapwings and the cry of cranes and the primordial wood-block rattle of the demoiselles. The demoiselles do not produce the clear bulging of their larger relatives, including *Grus vipio*—the sound that everybody listened for that night but failed to hear.

A dawn rain cleared off early. Not having to break camp, we departed quickly on a down-river crane hunt that would take us all the way to the China border. A few miles south, the track descended slowly toward a vast pale expanse of water called Khoyor Melkhit (Two Frog Lake), and as we drew near I saw an immense bird a half mile away, in the southeast corner of the lake, behind a gathering of gray herons. "See that big bird?" I yelled, as Jim yelled, "Stop!"

A superb adult white-naped crane stood transfixed by the dust of the braking vehicles, but in a moment it rose rapidly from the shore with hard upward licks of its nearly four-foot wings. This hard-flicking flight looked like threat display that one would expect to see in a breeding bird, and of course we hoped that somewhere nearby a mare and chicks awaited this majestic creature that flew away over the silver lake, its long neck and legs in silhouette against the eastern sun that rose up from the blue mountains of China.

Later in the day we proceeded north to some small lakes where Bold believed the lone white-naped might have headed, but in this season the lakes were dry, little more than shadows in the grass, haunted by nondescript shy birds, young Oriental plover. On our way back to camp, however, the vehicles flushed three cranes out of the marsh, and Harris, snatching up binoculars, called out, "Those aren't demoiselles!" A moment later he identified them as common, or Eurasian, cranes.

With Go, I walked across the plain to where the birds had settled down under the bluffs and was able to study through a telescope the first *Grus grus* seen on this journey. The white-crowned, black-throated, gray Eurasians were much larger than the demoiselles nearby, though considerably smaller than the *Grus vipio*. Like most cranes, the Eurasians are exceptionally wary, and the three birds soon rose up again and crossed over the marsh north of camp.

In a still dawn, two days later, a swan was calling from the lily lakes, solitary, mournful. We were breaking camp when two white-naped cranes flew in from the east, low against the valley rim, and, setting their wings, glided down toward the lake edge beyond the ruined temple not far from where, the previous afternoon, we had seen a pair with a gold chick—our first evidence of breeding *Grus vipio* on the Kerulen. Concerned (if these cranes were those we'd seen a day earlier) as to why they might have left their chick, I walked west to the ruin and on past the north end of the lake circling a bed of yellow blossoms at the marsh edge. Then, perhaps 200 yards ahead, the wary birds rose up from a low slough. With an intruder so close, one would have expected that they would flee, but they only alighted a little farther on, still within good range of my binoculars. They did not walk together, as pairs usually do, but crisscrossed each other's paths, entering the reeds from separate directions, coming out again, in what looked like an intense and systematic search, and all the while they uttered an odd call, a simple *ka-kuk*, sometimes only *ka*, which was the high note.

Unquestionable, or so it seemed, they were calling that chick, as if coaxing it to come in out of hiding. In a few minutes they rose again, only to glide across fifty yards of reeds to alight on a greensward where we'd seen the white-naped family the day before. There they resumed their intense search. Not once did their heads rise in alarm, nor did they pick and feed in nervousness. Still calling, they hunted in and out among the reeds until finally they did not emerge again.

If this was a second pair, as we now supposed, why were they hunting their chick so close to where the others had been yesterday? Unless the pairs no longer maintained territory, this made no sense. An unwelcome but inevitable explanation thrust itself forward—that the golden chick had been taken by some predator. Just as I'd concluded that this must be so—that the first known nesting of the white-naped in the Kerulen had ended in tragedy—a young crane emerged from the farther reeds and crossed quickly to the break in the reed wall into which the adult birds, still calling now and then, had disappeared.

I was so excited that I almost yelled, all the more so because this was not the previous afternoon's chick but a much larger one—larger, even, than the tall gray herons past which it ran on its way into the reeds. The gold-brown—a deeper color on the nape—was in this chick a scruffy brown and gray where new pin feathers had replaced the down. It seemed that the tragedy had been reversed and that this was a separate family, after all. I saw no more sign of the cranes, and they stopped calling.

At any rate, I had had a superb opportunity to study *Grus vipio* at close range and in good light. With its golden eye in a bright red and white head, and the shining white column of its nape in the fresh morning reeds extending down to the warm silver of the mantel, with throat and belly of a darker gray that seems to turn a lustrous black or reflects light, depending on its angle to the sun, it is surely the most "oriental" of the cranes, and the most striking.

Though we had established that *Grus vipio* bred in the Kerulen, the species was not there in any numbers, and so the expedition proceeded northward over the plateaus, across high plains, with steppe eagles and Mongolian gazelles—husky long-faced animals, uniform in tawny coloration but for the creamy rump.

On the Daurian Steppe, the grass thickened once again, with more wild-flowers—pale lavender aster and deep lavender thistle, night-blue spiky thistle, yellow lilies. The track wound eastward for some miles, then west to the

border town of Ereentsav and the great Lake Tari. Beyond lay grassland
and here, to the north, on a low hill of Siberia, stood a mixed stand of
conifers and poplar. Farther on, where the track ran beside barbed-wire fencing on the border, huge red machines of Siberian collectives were parked at
the ends of long grainfields separated by high windbreaks of tall poplars. A
pair of white-naped, rising from a reed pond in Mongolia, flew low across
the track and wire fencing and glided down into the Russian grain, where
they joined eleven of their own species, seventeen demoiselles, and two
Eurasian cranes, all in one flock.

To view them better, I left the car and ran ahead along the fence, as
a new flock of mixed Eurasians and white-naped crossed into Russia. The
white-naped alighted, but the Eurasians sailed on over the poplar
break to the adjoining field, where they joined two groups of feeding
demoiselles. Some of these cranes looked as black as crows, with a heavyset appearance, and when they raised white heads and necks, I yelled,
"Hooded!" The natural history of the hooded crane is very little known
(no nest was discovered until 1974). Like *Grus vipio*, it is one of Asia's
rarest cranes, with a population of perhaps 4,500, and I'd never seen one.
Harris, coming up the road, questioned the sighting, but a moment
later he said, "Hey! You're right!" From where we stood, we counted
forty demoiselles, sixteen Eurasians, thirteen white-naped, and thirteen
hoodeds.

I was elated, and Jim, too. With all his experience with wild cranes in
many countries, he had never before seen four species in a single sighting,
not even among the great crane assemblies at Poyang Lake in southeastern
China, where four crane species regularly spend the winter.

It was August now and we began our return west, heading up the valley of
the Uldz River. With low sad callings, two swans crossed the rolling hills
of grass on a day of dire cloud and restless wind that gave this landscape the
strong monotone cast of the Arctic tundra. Where the herds had been, the
river grass was cropped right to its banks, and in other stretches, in the
floodlands, the water was braided widely through the willows. In a grassy
place I found a hedgehog that rolled itself into a tight small ball. I turned it
gently, pulling apart a thorny indentation at the belly, revealing an alarmed
black eye and black shiny nose.

There were many swans at White Shears Lake, where Bold and
Tseveen had seen a Siberian crane in June of 1981. (With forty-five years of
wild-bird study between them, these two Mongolian ornithologists had

recorded the Siberian only three times.) But farther on, from a river bluff, we saw eight black storks, another rare species of this region. Like so many other species of the steppe, black storks can migrate as far as Africa; I'd last seen black storks in 1990, circling the Victoria Falls Hotel in the Zimbabwe sky. A pair of demoiselles fluttered and danced on the riverbank, strengthening the pair bond, even as we ran out into the marsh and captured and banded the nearly fledged chick of another pair. In doing so, we put to flight a pair of white-naped in a stubble field that seemed to circle back a bit too soon. When Chuka got a glimpse of the chick's head, we ran over and eventually found it—the tenth chick banded, and more important, the first white-naped.

Thinking a second chick might show itself, I crouched nearby after the first one was removed, hoping the adults' behavior would lead me to it. And soon they alighted not far away and crisscrossed in broad arcs all around me, until what was probably the male was within stoning distance, craning and peering over the field stubble, bill agape, gold eye flashing red reflections in the wild red face, soft gray wings silver in the stormy light. I had never seen a more splendid bird in all my life.

No chick showed itself; perhaps there was just one. When the chick was released, its parents kept their distance. They neither called nor moved toward it, and though it was peeping, it did not run to them, seeming to know it should hurry at once toward the cover of the river shrubs while they, almost aloofly, did the same.

The monsoon had arrived, and in steady rain we journeyed west each day, censusing and banding, checking sightings, pausing to view exciting birds such as the golden and white-tailed eagles, black storks, and great bustards. But the truly exciting thing that we were seeing was an extraordinary plenitude of white-naped cranes. A white-naped crane at the Dalt River, a tributary of the Uldz, then five more and two pairs of Eurasians, with one of the white-naped dancing, leaping, bounding, then two more *vipio* near the track, then two more, then ten pairs in one small valley, with seven more Eurasians—we were astonished. There were *vipio* every little way along this upper Uldz, but ten separate pairs in just one mile of river meadow was by far the greatest breeding concentration we had heard of anywhere.

The broad Uldz Valley narrows in low grassy hills, and soon the track left the Uldz headwaters behind, at a point perhaps 120 miles upstream from Lake Tari. Small trees scattered on the hills gathered into strands of conifers on the higher ridges, and the pines gradually descended as the track gained

altitude on its way west. In the south turned the silver gleam of a large
lake, and on a ridge a boy on horseback followed the black grass horizon,
bringing in a long slow herd of camels.

At the far end of a vast plain, under gray storm light and the shrouded
western sun, white clouds drifted over the high mountain ridges. After four
days' rain the plain was an interminable slough, and the vehicles lurched
drunkenly and churned and skidded through. A hare with brain transfixed
by headlights bounded ahead for minute after minute before jumping aside
into the grass. It was past ten when camp was made on a grassy hill, and it
was still raining.

A cold wind from the western steppe had driven back the weather, which
lay massed upon the hills on the far side of the Uldz Valley. A bright dawn
illuminated the tents, and the seesaw calls of demoiselles drifted across the
steaming grass, and dragging on clothes still dank and cold from the wet
crane hunts of the day before, everyone was cheerful. The track continued
west. A cinereous vulture crossed the sky, and a red-footed falcon plucked
a vole out of the grass not thirty yards from our tires. Soon we turned
north, departing the Uldz Valley and climbing to a ridge of dark pine and
pale larch, then descended to the Onon River, which flows north to the
Shilka in Siberia, an upper tributary of the great Amur that flows to the
Okhotsk Sea.

At the small town of Bayan Adraga (named for a holy mountain), we
turned west again, up the Onon, pausing in early afternoon to fish for supper in an Onon tributary, where Chuka brought in a small taimen,
a primitive salmonid that in the large Siberian rivers such as the Lena
may attain weights of well over one hundred pounds. I'd caught two
small pan fish on my fly rod, and was out on an open river bar, having
had a hard strike in a strong riffle, when a black storm rushed in over
the mountains—tremendous winds and lightning and thunder. Thunder
crashed all around as I thrashed across two tributary streams, fearful
of lightning.

I raced for cover just in time to reach the vehicles before an avalanche
of rain and hail, demonic in ferocity, left the whole river plain ghostly in a
strange white slush. The Mongolians were packing up even before the storm
relented, so concerned were they that this river plain would be hit in a flash
flood, and though we got out of there in time, the main track west was
already flooded out.

Heading cross-country, we climbed around the track, crossing a ridge of
conifers and descending once more to a high bluff where the Khurkh River

(pronounced backwards—*Hkruhk!*) joins the swift Onon, with a dramatic view over the confluence to the Hentiyn Mountains. Up there, Bold indicated, pointing northwest, was the sacred mountain of Burkhan Kaldun, which gave shelter from his enemies to the young Temüjin, who later took the name of Genghis Khan.

Bold wasted not a moment getting ready to go night-fishing for the taimen, which he hoped to entice with a crude furry lure the size of a young muskrat—a simulated mouse, said Chuka (who was similarly equipped), that is dragged slowly across the surface like a struggling swimmer caught in the swift current. But the rain continued and the night was very dark, and we caught nothing but heavy river weeds torn out by storm, and we gave it up until 3:15 A.M., when I was awakened by Bold's ferocious old alarm and staggered down in darkness to my pre-selected spot only to find it was a foot beneath the rising river. Also, the water was roily, and under the bluff there was no place to strip my line, let alone backcast, and I retreated to the tent, waiting for light.

In a dark daybreak, employing a grotesque rig of my own devising—a saltwater spoon heavy enough, when hurled out shot-put fashion, to drag the fly line coiled upon my toes up through the guides and out over the river—I managed to foul-hook a taimen that missed its crude swipe at my lure in the murky water and got snared behind the gills instead, making it hard to turn its head with my light equipment. Drawn to the scene by the great thrashing, Chuka dropped his rod and raced to help me even as I shouted at him to stand back, but he was right there in the shallows when finally I brought the fish in close, and he heaved it ashore joyfully with both hands, then up the steep bank onto the grassy bluff.

The dull metallic greeny-gray fish with its red tail and pelvic fins measured out at forty inches long—a fair fish in the high-altitude waters of Mongolia—and proved to be a delicious change from old black goat, which had not improved in its long days of unrefrigerated travel. Nonetheless, the redoubtable Bold (whose name signifies "hard," as in hard steel) was still chiseling stray meat bits from the smoke-blackened skull with its staring teeth and eating them from the point of his old pocket knife.

We headed west again under gray skies and intermittent rain, wondering if, on such muddy tracks, we would ever reach Ulan Bator, still more than two hundred miles away. Apart from the weather, the small settlements we were passing had no fuel, not even enough for the tractors that had to cut the hay for the coming winter. That night we spent two hours scrounging precious fuel in Dr. Tseveenmyadag's brother's village of Omnödelger.

More and more as we penetrated this back-country, we discovered that the white-naped crane, where protected by the benevolence or superstition of the local herders, did not require large and pristine breeding territories but would nest in a relatively small marsh and within sight of human habitation—in one place, with ten *gers* on three sides of a small lake with limited reed cover for the nest.

From a rock outcrop on the valley ridge, I could see up and down the long and lake-eyed valley of the Khurkh, with its white *gers*, far herds, long-maned horses reflected in still pools, slow camel trains, and grass and grass and grass stretching away in a bright sun to the purple heights of the storm-shrouded mountains. Beyond the knot of my colleagues surrounding the two chicks they were banding, the distressed parents pretended to feed, straightening long white-naped necks every few moments to peer over the reeds at the human beings, then lifting to flap and glide and alight and peer again, from another angle.

Up on the ridge, I felt a little sad. Already I was beginning to miss Mongolia, knowing how unlikely it was that I would ever travel here again. Our expedition was coming to an end, and perhaps these white-naped were the last that we would see, for we were far from the Uldz Valley now and close to the western end of their range (*Grus vipio* had been observed west of Ulan Bator, but to date no breeding pairs have been verified). But toward six that evening, in a broad valley of shallow lakes, there were twenty-six of them in sight at once, the largest group observed on the whole journey, and one that seemed to justify our conclusion that this far western region of the Daurian Steppe, in the upper valleys of the Uldz and Khurkh, was the very heart of the heartland of this bird.

One of the pairs had a large chick, and Tseveen went straight to the nearest *ger*. Within minutes he was galloping with a young horseman the two or three miles to where the chick, much too big to hide and very nearly fledged, set off on a wild flapping run across a reed bed and out onto the plain. Jouncing and swerving at dangerous speed, our car cut off its escape and we jumped out to surround it.

In the end, Jim Harris bravely grasped it by the neck and subdued it without harming the crane or himself, though he was nicked by its sharp toenails in the process. "This is the first dangerous bird we've handled," he said later, though he was not thinking about toenails at the time. Already this bird was a beautiful fawn color where the rich white nape was going to be, and at this size, so nearly fledged, its bill had hardened; given a chance, it would aim for the eyes. When finally it was released, the fierce young creature did not flee but turned around and advanced upon Go Fujita, then fluttered upward, pecking toward his face. Not until it drove him back did the young crane turn away and, in no great hurry, stalk away across the plain, crossly shuffling its new feathers, shaking the bright green ring around its leg.

—11—
Science Is. . . .

Ingrid Newkirk

I f we listen to our souls, we'll abandon Science. Official Science. The Science that believes it wrote the songs. The Science that conveniently forgets how shamans and chimpanzees have used natural anesthetics and antibiotics, birds have painted wounds with healing plants and therapeutic muds, and even plants have used effective insecticides since time began. The Science that pretends the wisdom of tribal peoples is far too primitive to be important and thinks it a bad dream that witch doctors in Borneo were successfully trepanning patients long before Western tinkerers picked up their first scalpel.

I'm not talking, or course, about the sort of empathic, non-invasive, sympathetic Goodall-Galdikas-Fossey studies which conjure up new respect for life in all its diverse forms, but, rather, rigid, mechanistic Science that, at a minimum, needs heavy scrutiny but never gets it: witness diethylstilbestrol, bovine growth hormone, and Depo-Provera.

Science is a "discipline" incongruously out of control. Like Caligula, it has become sufficiently self-enamored to declare itself a god, dismissing those things which connect us most to the earth and to each other—love, other emotions, our intuitions—with the wave of a plastic-gloved hand.

Science deserves to be indicted for betraying our trust. Its victims include my father on a golf course, confidently pumping out clouds of DDT from a tin onto our dachshund to kill fleas ("perfectly safe"); his aerial equivalent "dusting" (what an understatement) fields with a blanket of insecticide; and tumor-ridden fishes with bloated eyes floating belly up in the warm run-off from power plants.

Science is the evacuation of Love Canal (Science must have hated the name), red ant poisons contaminating the soil in southern playgrounds, trains full of nuclear waste cruising the countryside anxious to derail and get it over with because they have nowhere else to go, rats screaming as their gonads are removed in a sex experiment in the basement of New York University, and pigeons languishing upside down in empty water beakers for the convenience of the University of Pennsylvania psychology professors.

Science is the Tuskegee syphilis experiments and the atom bomb; its emissaries men in white coats on television telling us which cigarettes soothe

the throat and the learned founder of Yerkes Regional Primate Research Center declaring in a lecture to Harvard that mixing with Negroes would lower the median intelligence in the United States of America.

Science is a bully. It picks on things (and beings) smaller and less powerful than itself or on huge inanimate targets it bets won't strike back, like the ecosystem. Like most bullies, it is also a coward which screams like a baby when someone threatens to do to it what it does, with impunity, to others: peer closely, take it apart, analyze it, try to establish its worth, or lack thereof, in relation to its usefulness to others. "Get away, don't touch me," bellows the Emperor Without Clothes as it runs its own hands over total strangers against their will, turns a thing (or a being) upside down, smashes it, stares at the bits, then, on a whim, puts the parts back sideways "just to see what happens."

Science is pompous. It demands that no restrictions whatsoever be placed on its quest for knowledge, no matter how trivial or inapplicable or grotesque its methods of extracting a fact or a factor may be. If it finds its back against the wall, it invents the myth of "self-policing." While denying "interests" to all others (how hard won were patients' basic rights to informed consent) it declares its right to remain off-limits in its ivory towers and behind its key-card-equipped steel doors.

Science, in its arrogance, feels compelled to rearrange nature. It believes it can simply adjust the knob here or pop a baboon organ in over there and the world will toast its adventure with a plastic bottle of machine-extracted, genetically engineered champagne grapes. Of course people do cheer initially, especially if they have related investments which are putting the nuclear family's offspring through Neurophysiology 101. It's only later that the crunch comes and, fingers crossed, we'll be in Forest Lawn by then.

That wonderful medical heretic, Dr. Robert Mendelsohn (I might forgive Science if it had ever achieved anything useful, like bringing back the good dead), said of Science's best-loved son, Medical Science, "It has convinced itself that Life is a disease that requires treatment from the moment of birth until death." If I mount one more talk show platform to hear an experimentalist declare, "We all owe our lives to Medical Science," my screams will shatter glass. Or perhaps I shall quietly ask, in memory of Mendelsohn, whether the speaker is asserting that God or nature, take your pick, was at fault for not putting a zipper in my mother's stomach so I might be brought into this world in a way convenient enough to allow her doctor to complete the eighteenth hole.

Science adores artifice as much as it despises nature. It dismisses as "anthropomorphic" any observations of the intelligence and sentience of beings it prefers to think of as unimportant machines. It makes its nest in the cold, unnatural setting of the windowless, temperature-controlled laboratory with its tube lighting and cold metal cages where it can privately manipulate its toy-of-the-moment until seduced by a shinier toy. In its private chambers it invents its own isolated language, designed not to elucidate but to obscure the real and nasty—cries of tormented cats become "vocalizations," an electric shock becomes a "negative stimulus," and when birds open-mouth kiss during erotic play, that becomes "false-feeding."

Science is happiest knowing *nothing*, except that Science is great and mustn't be disturbed even during the laboratory equivalent of happy hour—when the federal grant checks arrive. It refuses to recognize a truth until it can be perfectly and repeatedly quantified, thus effectively eliminating the possibility of harmony and rendering unreal the unmeasurable: pain, grief, and altruism, for example. "To comprehend the organs of the horse is not to comprehend the horse itself" is wisdom Science can't bear to think about without giving itself a migraine. It deals only in "organs," of one kind or another, removing them from their "host," all the better to play with, and adding or subtracting "variables" the way a chef might experiment with decorative parsley.

Science is a precocious, maladjusted child raised without ever going outdoors to experience life. Not knowing the value of a dollar, it requisitions millions with which to spend years deciding if cocaine-addicted monkeys self-mutilate, while outside in the real world human addicts are left to rot. (Yes, you can make the same argument about Art, but I'll make it: that money is precious, it's needed by—choose one—homeless schizophrenics, Somalian refugees, migrant grapepickers.) It doesn't even possess the common sense of the fruit flies whose brains it casually scrambles. It feeds tequila to guppies to see if they "demonstrate lack of coordinated motor function" when swimming drunk; it fails to grasp such a fundamental concept as maternal love, although it incarcerates wild-caught rhesus monkeys for years in expensively constructed deprivation chambers and writes copious notes every quarter hour on their misery. Even the "result" of its research—"inconclusive"—is calculated to deny the obvious and to draw more negotiable lettuce through its mailbox.

Science has committed too many grotesqueries to be considered amusing. Its high priests listen to the linnet trill his amazing love song to his life-mate, then capture him, cage him, extract portions of his brain, castrate him, plant wires in his head, so that he can no longer fly or love or live, and record how these molestations affect his calls. I could tell you more, but isn't this enough all by itself? If Science did nothing worse than deliberately destroy one little bird to satisfy its curiosity, that is all that should ever be needed to label it unfit to exist in society. Who can criticize any moral standard, yet permit experimentalists to move freely among us, buying their groceries in our neighborhood stores, washing their cars and thinking their ugly thoughts on the streets where our children play?

If we are to preserve what little Science has yet to ruin, we should remove the experimentalists from polite company, strip them of their living toys, and send them into the community to do productive deeds, like planting organic vegetables in window boxes in the Bronx or teaching remedial reading using books like *Silent Spring* and *Born Free*. Considering the amount of destructive exploitation attributable to Science since the Industrial Revolution, I'd be happy if its disciples were paid a stipend to sit under palm trees and play acoustic guitars for the rest of their lives.

Nature would be the better for it.

—12—
Understanding the Great Mystery

Joseph Bruchac

In 1992, while visiting Baffin Island to talk with Inuk elders about hunting and its relationship to the balance of all things, an old hunter named Akaka told me a story of an old walrus. That old walrus offered itself to a young hunter as the man paddled his quyak through the ice floes where there were many walrus, trying to decide which animal to strike with his harpoon.

"I want a drink of water," the old walrus said. "Give me a drink of water."

But instead of killing the animal, the hunter laughed at it.

"Your tusks are too dark. You are too thin and old for me to kill."

As soon as the hunter said this, not only the old walrus, but all of the other walruses slid from the ice into the sea. None of them came back up and the hunter went back home empty-handed.

I think of that story often, for it exemplifies so much about the way Native people see the world and interact with that great mysterious force which those raised in European culture called "nature."

In European thought, nature and "man" are separate. The Old Testament, in English translation, at least, says that humans have been given dominion over nature. Further, The New Testament includes passages which equate "the world" with "the devil." That which is spiritual is often defined as being above or apart from nature. In much of traditional Western thought, the natural world is an adversary, a mindless, spiritless thing to be controlled, subdued, and used. This can be seen in virtually all of European literature, past and present, especially the folk and "fairy" stories in which the natural world is a dangerous forest, a desert wilderness where monsters dwell and where the devil holds sway. When Hansel and Gretel are lost in the forest they find themselves the captives of a cannibalistic witch. In the story of "Little Red Riding Hood," a wolf comes out of the forest to eat the little girl's grandmother and threaten the child herself. At the appearance of a woodsman, a person whose livelihood destroys the forest, the heroine is saved and the wolf killed.

It is only within the last hundred years that the idea of wilderness as a positive thing has crept into European consciousness with the creation of a conservation ethic. The influence of Thoreau is important. But few seem to know that the New England transcendentalist's ideas were shaped not only by Asian thought but also by the American Indian vision of spirit in nature.[1]

Let us turn back to that Inuk story and look at the lessons that it teaches, for they are crucial to understanding that which cannot be truly understood, that cosmic, spiritual force of Creation known by many different names, including *orenda* in the Iroquois language, *manitou* in Anhisnabe, *wakan* in Lakota. In the language of my own Abenaki ancestors it is spoken of as *manido* and sometimes as *nwaskw*. One of the terms we use to refer to the Creator of all things is Ktsi Nwaskw, "Great Mystery."

In the Native cultures of North America, the work of the traditional hunter is not only vital for the survival of the hunter's family, it is also filled with danger. All life is sacred and so when a hunter goes out to deliberately take the life of another being, that hunter is doing a very serious thing. One's mind and one's spirit must be in the right relationship or one will fail in hunting. An unprepared hunter might even be injured or killed. Among the Abenaki people, going to hunt for an animal such as a bear might require ceremonial purification in the sweat lodge before setting out. Unlike hunting in contemporary terms, where there is hunter and prey, engaged in a purely physical process, in North American Native thought hunting is a process that involves cooperation between the spirits of hunter and hunted. It is a sacred undertaking. In the Pueblos of the Southwest, special songs are often sung to the spirits of deer before going to hunt them and pollen and bits of turquoise may still be placed in the footprints of deer when tracking them. The spirit of the hunter speaks to the spirit of the hunted animal asking permission to kill it. The necessity of the hunt is explained—not for sport or personal gain, but to help the hunter's people survive by providing them with food and clothing and all the things an animal's body may provide.

If the hunter explains his purpose clearly, then an animal may choose to sacrifice its body, though its spirit will survive. I know a Seneca story in which a hunter shoots an arrow at a bear. That bear allows the arrow to strike it. Then the bear continues on its way, after throwing down a skin bag full of meat. That skin bag full of meat is the bear's physical body, but its spirit continues to survive unchanged by death. However, if the hunter does not treat the body of the animal with respect, the spirit will observe this disrespect and in the future that hunter will not be successful. Part of the respect may be an exchange of gifts. The animal gives its body to the people and the hunter may offer something to the animal's spirit. When an Abenaki hunter kills a bear, for example, he may offer tobacco.

In the case of the walrus in the Inuk story, it is said that sea animals such as seals and walrus are thirsty for fresh water. When killing a seal or a walrus, a traditional Inuk hunter melts some snow in his own mouth and allows that water to fall from his own mouth into the mouth of the animal, giving it a drink. When the Inuk hunter in the story refused to accept the great gift offered him by the old walrus, which said it wanted a drink of water, thus indicating its willingness to die, he showed great disrespect. The lessons of this story are many. Our lives are not merely physical and we cannot measure things by their appearance. Spirit is in all things. We must show respect for all things and accept the gifts of spirit in the proper way. If we fail to do this, then we will suffer.

That lack of understanding of the spiritual nature of all life, of the great mysterious spirit in nature which we must respect, has been at the root of the current environmental crisis we now face. When we think we own nature, as one might own an article of clothing, we begin to see it as something we can wear or simply discard when it is no longer useful. When we see nature as a thing, without spirit, then our actions towards nature are seen as having no spiritual consequences.

But the Native way is very different. The stewardship of human beings is not as owners, but as partners with many other beings, such as the animals. The animals are recognized not only as spiritual beings but, in some ways, as beings wiser than humans. Unlike humans, they do not forget the right way to behave. A bear never forgets that it is a bear, yet human beings often forget what a human must do. Humans forget to take care of their families and forget to show respect to other things. They become confused because of material possessions and power.

This potential for confusion was true for Native people long before the coming of Europeans (though Europeans have raised the art of spiritual confusion to a new level!) That is why the traditional teachings remain so important. They remind human beings how to take care; the old stories, for example, remind people of the right way to behave. The Cherokee tell the story of Awi Usdi, the Little Deer, who makes sure that hunters behave properly. If a hunter should kill a mother animal with young ones or kill more animals than are needed, then Little Deer will visit that hunter in the night and inflict him with rheumatism so that he will be too stiff to hunt.

When hunting is done, as it has been done in the West, for sport or for profit, then no concept of physical or spiritual balance rules the hunter's actions.

If there is a certain number of buffalo, then—to a Western hunter—that is the number which can be killed. The result of such market hunting in the nineteenth century was the extermination of multiple populations of animals.

A similar attitude has been taken by Europeans towards the forests of the Americas, seeing them as wilderness to be cleared or as logs to be harvested. Though the concept of the sacred grove existed in ancient Europe, it has not been carried over into the Christian era. But the Native view is quite different. Chan K'in is the spiritual elder of the Lacandon Mayan people of Mexico. The lives of his people depend upon the rain forest, which is shrinking further each year, but Chan K'in, who is well over 110 years old, sees it in larger, more spiritually interconnected terms. "Each time a great tree falls," he says, "a star falls from the sky. Before we cut a tree, we must ask the guardian of that tree. We must ask the guardian of that star." When I spoke with Chan K'in recently he told me of his fears for all living things if we destroy all the forest. "The great cold and the great dark will come upon us," he says.

Like the animals, the plants have spirits. Throughout North America, whenever plants are harvested, it is done with an awareness of the life and spirit of the plant. To gather medicine plants, for example, one must be in the right frame of mind. "If your spirit is wrong when you go out to gather medicine," an Onondaga elder told me, "then those plants will hide. You won't be able to find them at all." The Iroquois people refer to the primary food plants, the Corn, Beans, and Squash, as the Three Sisters, those who sustain us. When planting those seeds, working in the gardens, and harvesting, there are old songs and ceremonies which honor the spirits of the Corn, the Squash, and the Beans. It may be said, quite literally, that there is no difference in Native North America, between planting and prayer.

I have often heard it said by Native elders of different nations that "We Indians have no religion." This is a statement made because of the great knowledge which Native people have of Western religion after five hundred years of being proselytized. Native people have been viewed at various times as beings with no spirit, "like animals" (a simile which carries some special irony for Native people), as devil worshippers, or as empty spiritual vessels ready to be filled with whatever brand of Christianity is being served up by the missionary of the day—Baptist, Catholic, Protestant, Seventh Day Adventist, Mormon. A few years back, a friend of mine named Swift Eagle had a little Native American jewelry shop on the main street in Saratoga Springs, New York. As I sat with him one summer afternoon, he was visited by no fewer than five different missionaries, all wanting to convert him from his heathen ways to their *true* religion.

Red Jacket, the great Seneca orator, made a famous response to a member of the Boston Missionary Society at a meeting called in 1805. "Brother," Red Jacket said, after being lectured at length about the follies of Indian paganism, "you say there is but one way to worship and serve the Great Spirit; if there is but one religion, why do you white people differ so much about it?" Red Jacket also showed his own knowledge of Christianity even further by pointing out that "If you white people murdered 'the Savior,' make it up among yourselves. We had nothing to do with it. If he had come among us we should have treated him better."

Red Jacket might also have pointed out that the Iroquois and other Native people did not see the natural world as being evil or at odds with spirituality, but that the presence of *orenda*—the sacred—the presence of the Great Spirit is to be found in all things. If religion is defined as something which separates human beings from the plants, the animals, the stones, and places them in a position of dominance over, or opposition to, nature, then "religion" is something that was indeed foreign to the Native peoples of North America. Further, if religion and one's spiritual life is relegated only to what goes on inside a building one day in every seven, then that "religion" was something that Indians neither needed nor wanted. Red Jacket went on to say that he was aware of the fact that the missionary had been preaching to the Christian neighbors of the Senecas. Yet, being Christian did not seem to prevent them from mistreatment of the Indians. "We will wait a while," Red Jacket said, "and see what affect your preaching has upon them. If we find that it does them good and makes them more honest and less disposed to cheat Indians, we will then consider again what you said."

The idea of the sacred being everywhere is expressed in many ways by Native people. Long Standing Bear Chief, a Blackfoot elder, published a book call *Ni Kso Ko Wa* in 1992. It explains Blackfoot spirituality and traditions and contains this simple and eloquent statement about sacred places:

> Wherever you are doing something ceremonial, that place becomes sacred. The entire earth and everything about it is sacred, not one place more than the other.

Because of the powerful sacred nature of Creation, then, there is ideally no distance between human beings and the earth. It is said among the Mohawk that the faces of our children not yet born are just there, under the earth. It is also commonly said throughout the continent that the very old and the very young are especially blessed because they have either just come from the earth or are just about to return to the earth. Both ages are very close to the Great Mystery.

When one begins to understand the Great Mystery, to know (in a way which transcends conventional knowing) that our human spirits are part of a great circle of spirit, then that understanding must also translate into action. We begin with thanks. An Onondage elder told me that he reminds himself to be grateful every day for everything he is given. He says a prayer of thanksgiving each time he drinks a glass of water. If we were all to be so thankful and so aware of the powerful, mysterious, spiritual gift which water is, would we not be more likely to keep it clean for future generations? An elderly Mohegan woman named Fidelia Fielding kept, in the early part of the century, a journal in her own language. Some years ago I worked on a translation of her works and I was moved by their simple eloquence. Each morning she recorded her thanks for simple things, being able to get up from her bed, being able to stand. "*M'undu wi go*," she wrote, "The Creator is good. *Ni ya yo*. It is so."

There is so much to be thankful for and so many gifts which we fail to acknowledge. Yet those simple words of Fidelia Fielding, whose Indian name was Djits Budunacu, "Bird Which is Flying," offer a good place to begin. With such acknowledgment and thanks we begin to walk a balanced path of caring and caretaking, a path which leads us back into the circle of spirit. And in that circle we may begin to understand the Great Mystery.[2]

Two Poems
"Singing to the Dead" and "Oh Great Spirit"

Deena Metzger

Singing to the Dead

for Brenda Peterson

A layer of white chalk, hundreds of feet deep, lies under England, Europe, and the Mid-East. Under the dark earth, under the moist and pungent humus, lies a hidden strata of white bones, countless fossil shells of tiny animals that once lived in the ancient seas. And here and there caught in the white chalk are other creatures, ancestors of those we know, white bones of ancient crocodiles, wolves who the spirits say became whales or whales which once were men, held fast in the subterranean layers laid down millions of years ago.

The spirits of the animals lie like strata of milky knowledge distilled from crushed bones at the depths of our souls.

There was a poet whom I love, who lived his life underwater or among the ravin wolves or among the sea birds. I was lonely for him, so we went north together to drink the elixir of that chalky light and feed it to the dead.

I had forgotten to tell the dead a story and stories are what we must feed to the dead.

They come to the edge of the story, the way the animals gather at a clearing, grazing at the edge of a circle of grass or drinking at the rounded corners of the stream or swimming in the edges of centrifugal whirlpools of dark waters.

We stand at the other limits, at the other side of the stream or where the trees form a safe border for the grass and gradually we slide toward the dead, ease ourselves along the white roots toward the locus of all beginnings. Or we circle down the light to the shining deadly point of time. It is high moon. We are illuminated, living and dead, in the luminescent shadows, in the lunar seas and the silvery light.

We make a fire for the dead from dew and light. We make a fire of language and vision. We invite them to warm their hands at this fire of stories and to drink these potions of sea water and poems.

We started with offerings. We fed bread to the gulls. The clouds came in hungry too and it almost rained. We were grateful for the manna of rain and everything which is white.

I had asked the dead what they longed to hear and they said they were lonely for the animals with whom they used to sing.

We went to the penitentiary of the wolves and the prisons of the walruses and whales. The white wolves sang to us from their small compounds where they are enclosed.

Too injured to return to the wild, ignorant or trained not to find their own meat and habitat, they are fed here at the edge of winter, deprived of hunger and lonely wandering, of the wary life at the edges of the woods, even of the fear of human kind. We threw back our heads, the notes threaded through the confines of gates and barbed wire until we were tied into the harmony of the howls.

Behind glass walls, the Beluga swam around in the small portion of sea, contained far from the gulls and the sonar deep. We pressed our hands and eyes against the glass and they came swirling by.

How lovers separated by a window, each one following the other, trace the patterns of their loving in a careful dance. How longing has a curve to it, a certain grace that dips and eddies in a desperate wave.

What I must tell the dead is that I sang to them.

The whales came swooping through the water and humming pressed out between our lips, slipped through the glass and into the water. We could see it go its salty way and we followed the gleaming song toward the white creatures. The white whales came toward us, rubbing against the glass, playing against our hands, their shimmering sonar sounding through our bones as our hum entered their bodies. Song now undulating along their sleek skin, smoothing the ragged scars, white flippers, and the gleaming tails.

At sunset, the walrus climbed up the ledge by his little pool and pressed his mouth and stiff beard of sea weed over the small fence against my hand. I reached to press my mouth against his mouth to thank him for the kiss.

These are the stories for the dead. The spirits of the animals form a white chalk layer at the depths of our lives. Fossils of the great and small, miracles in white bones, ground to a milky dust, pool beneath our feet. We drill a small hole into the center of the earth, describe a careful descent along the luminous circle where the living and the dead gather around the well of the moon and the singing animals offer up their white knowledge from the deep.

The poet whom I love lived his life underwater or among the ravin wolves or among the sea birds. I was lonely for him. I called out through the sediment of bone beneath my heart. When we are lonely, the choruses of the dead come toward us in the clearing and the creatures of the wild rise up in song.

Oh Great Spirit

I n the name of Raven. In the name of Wolf. In the name of Whale. Who have taught us. Who have guided us. Who have sustained us. Who have healed us.

Please heal the animals.

In the name of Raven. In the name of Wolf. In the name of Whale. In the name of Snake. Whom we have slaughtered. Whom we have feared. Whom we have caged. Whom we have persecuted. Whom we have slandered. Whom we have cursed. Whom we have tortured.

Please protect the animals.

In the name of Raven. In the name of Wolf. In the name of Whale. In the name of Snake. Whose habitat we have stolen. Whose territory we have plundered. Whose feeding grounds we have paved or netted. Whose domain we have poisoned. Whose food we have appropriated. Whose young we have killed. Whose lives and ways of life we threaten.

Please restore the animals.

In the name of Raven. In the name of Wolf. In the name of Whale. In the name of Snake.

Forgive us. Have mercy. May they return. Not as a resurrection, but as living beings. Here. On earth, on this earth that is also theirs.

Oh Great Spirit. Please heal the animals. Please protect the animals. Please restore the animals.

So our lives may also be healed. So our souls may also return. So our spirits may also be restored.

Oh spirit of Raven. Oh spirit of Wolf. Oh spirit of Whale. Oh spirit of Snake.

Teach us, again, how to live.

—14—
Undressing the Bear

Terry Tempest Williams

He came home from the war and shot a bear. He had been part of the Tenth Mountain Division that fought on Mount Belvedere in Italy during World War II. When he returned home to Wyoming, he could hardly wait to get back to the wilderness. It was fall, the hunting season. He would enact the ritual of man against animal once again. A black bear crossed the meadow. The man fixed his scope on the bear and pulled the trigger. The bear screamed. He brought down his rifle and found himself shaking. This had never happened before. He walked over to the warm beast, now dead, and placed his hand on its shoulder. Setting his gun down, he pulled out his buck knife and began skinning the bear that he would pack out on his horse. As he pulled the fur coat away from the muscle, down the breasts and over the swell of the hips, he suddenly stopped. This was not a bear. It was a woman.

Another bear story: There is a woman who travels by sled dogs in Alaska. On one of her journeys through the interior, she stopped to visit an old friend, a Koyukon man. They spoke for some time about the old ways of his people. She listened until it was time for her to go. As she was harnessing her dogs, he offered one piece of advice.

"If you should run into Bear, lift up your parka and show him you are a woman."

And another: I have a friend who manages a bookstore. A regular customer dropped by to browse. They began sharing stories, which led to a discussion of dreams. My friend shared hers.

"I dreamed I was in Yellowstone. A grizzly, upright, was walking toward me. Frightened at first, I began to pull away, when suddenly a mantle of calm came over me. I walked toward the bear and we embraced."

The man across the counter listened, and then said matter-of-factly, "Get over it."

Why? Why should we give up the dream of embracing the bear? And what do these bear stories have to do with writing about the natural world? For me, it has everything to do with undressing, exposing, and embracing the feminine.

I see the feminine defined as a reconnection to the self, a commitment to the wildness within—our instincts, our capacity to create and destroy; our hunger for connection as well as sovereignty, interdependence, and independence, at once. We are taught not to trust our own experience. The feminine teaches us experience is our way back home, the psychic bridge that spans rational and intuitive waters. To embrace the feminine is to embrace paradox. Paradox preserves mystery, and mystery inspires belief.

I believe in the power of Bear.

The feminine has long been linked to the bear through mythology. The Greek goddess, Artemis, whose name means "bear," embodies the wisdom of the wild. Christine Downing, in her book, *The Goddess: Mythological Images of the Feminine*, describes her as "the one who knows each tree by its bark or leaf of fruit, each beast by its footprint or spoor, each bird by its plumage or call or nest. . . ."

It is Artemis, perhaps originally a Cretan goddess of fertility, who denounces the world of patriarchy, demanding chastity from her female attendants. Callisto, having violated her virginity and becoming pregnant, is transformed into the She-Bear of the night sky by Artemis. Other mythical accounts credit Artemis herself as Ursa Major, ruler of the heavens and protectress of the Pole Star or *axis mundi*.

I saw Ursa Major presiding over Dark Canyon in the remote corner of southeastern Utah. She climbed the desert sky as a jeweled bear following her tracks around the North Star, as she does year after year, honoring the power of seasonal renewal.

At dawn, the sky bear disappeared and I found myself walking down the canyon. Three years ago, this pilgrimage had been aborted. I fell. Head to stone, I rolled down the steep talus slope stopped only by the grace of a sandstone boulder precariously perched at a forty-five degree angle. When I stood up, it was a bloody red landscape. Placing my hand on my forehead, I felt along the three-inch tear of skin down to the boney plate of my skull. I had opened my third eye. Unknowingly, this was what I had come for. It had only been a few months since the death of my mother. I had been unable to cry. On this day, I did.

Now scarred by experience, I returned to Dark Canyon determined to complete my descent into the heart of the desert. Although I had fears of falling again, a different woman inhabited my body. There had been a deepening of self through time. My mother's death had become part of me. She had always worn a small silver bear fetish around her neck to keep her safe. Before she died, she took off the bear and placed it in my hand. I wore it on this trip.

In canyon country you pick your own path. Walking in wilderness becomes a meditation. I followed a small drainage up one of the benches. Lithic scatter was everywhere, evidence of Anasazi culture, a thousand years past. I believed the flakes of chert and obsidian would lead me to ruins. I walked intuitively. A smell of cut wood seized me. I looked up. Before me stood a lightning-struck tree blown apart by the force of the bolt. A fallout of wood chips littered the land in a hundred-foot radius. The pinon pine was still smoldering.

My companion, who came to the burning tree by way of another route, picked up a piece of the charred wood, sacred to the Hopi, and began carving a bullroarer. As he whirled it above our heads on twisted cordage, it wailed in low, deep tones. Rain began—female rain falling gently, softly, as a fine mist across the desert.

Hours later, we made camp. All at once, we heard a roar up canyon. Thunder? Too sustained. Jets overhead? A clear sky above. A peculiar organic smell reached us on the wind. We got the message. Flushed with fear, we ran to high ground. Suddenly, a ten-foot wall of water came storming down the canyon filling the empty streambed. If the flood had struck earlier, when we were hiking in the narrows, we would have been swept away like the cottonwood trees it was now carrying. We watched the muddy river as though it were a parade, continually inching back as the water eroded the earth beneath our feet.

That night, a lunar rainbow arched over Dark Canyon like a pathway of souls. I had heard the Navajos speak of them for years, never knowing if such magic could exist. It was a sweep of stardust within pastel bands of light—pink, lavender, yellow, and blue. And I felt the presence of angels, even my mother, her wings spread above me like a hovering dove.

In these moments, I felt innocent and wild, privy to secrets and gifts exchanged only in nature. I was the tree, split open by change. I was the flood, bursting through grief. I was the rainbow at night, dancing in darkness. Hands on the earth, I closed my eyes and remembered where the source of my power lies. My connection to the natural world is my connection to self: erotic, mysterious, and whole.

The next morning, I walked to the edge of the wash, shed my clothes, and bathed in pumpkin-colored water. It was to be one of the last warm days of autumn. Standing naked in the sand, I noticed bear tracks. Bending down, I gently placed my right hand inside the fresh paw print.

Women and bears.

Marion Engel, in her novel, *Bear*, portrays a woman and a bear in an erotics of place. It doesn't matter whether the bear is seen as male or female. The relationship between the two is sensual, wild.

The woman says, "Bear, take me to the bottom of the ocean with you, Bear swim with me, Bear, put your arms around me, enclose me, swim, down, down, down, with me."

"Bear," she says suddenly, "come dance with me."

They make love. Afterwards, "She felt pain, but it was a dear sweet pain that belonged not to mental suffering, but to the earth."

I have felt the "dear sweet pain" that belongs to the earth, pain that arises from a recognition of beauty, pain we hold when we remember what we are connected to and the delicacy of our relations. It is this tenderness born out of a connection to place that fuels my writing. Writing becomes an act of compassion toward life, the life we so often refuse to see because if we look too closely or feel too deeply, there may be no end to our suffering. But words empower us, move us beyond our suffering, and set us free. This is the "sorcery of literature." We are healed by our stories. To articulate what we know in our hearts is never easy. Solitude is required in order to listen, courage in order to speak.

By undressing, exposing, and embracing the bear, we undress, expose, and embrace our authentic selves. Stripped free from society's oughts and shoulds, we emerge as emancipated beings. The bear is free to roam.

If we choose to follow the bear, we will be saved from a distractive and domesticated life. The bear becomes our mentor. We must journey out, so that we might journey in. The bear mother holds the secrets of regeneration within her body. She enters the earth before snowfall and dreams herself through winter, emerging in spring with young by her side. She not only survives the barren months, she gives birth. She is the caretaker of the unseen world.

As a writer and a woman with obligations to both family and community, I have tried to adopt this ritual in the balancing of a public and private life. We are at home in the deserts and mountains, as well as our dens. Above ground in the abundance of spring and summer, I am available. Below ground in the deepening of autumn and winter, I am not. I need hibernation in order to create.

We are creatures of paradox, women and bears, two animals that are enormously unpredictable, hence our mystery. Perhaps the fear of bears and the fear of women lies in our refusal to be tamed, the impulses we arouse and the forces we represent.

Last spring, our family was in Yellowstone. We were hiking along Pelican Creek, which separated us from an island of lodgepole pines. All at once, a dark form stood in front of the forest on a patch of snow. It was a grizzly, behind her, two cubs. Suddenly, the sow turned and bolted through the trees. A female elk crashed through the timber to the other side of the clearing, stopped, and swung back toward the bear. Within seconds, the grizzly emerged with an elk calf secure in the grip of her jaws. The sow shook the yearling violently buy the nape of its neck, threw it down, clamped her claws on its shoulders, and began tearing the flesh back from the bones with her teeth. The cow elk, only a few feet away, watched the sow devour her calf. She pawed the earth desperately with her front hooves, but the bear was oblivious. Blood dripped from the sow's muzzle. The cubs stood by their mother, who eventually turned the carcass over to them. Two hours passed. The sow buried the calf for a later meal, then slept on top of the mound with a paw on each cub. It was not until then that the elk crossed the river in retreat.

We are capable of harboring both these responses to life in the relentless power of love. As women connected to earth, we are nurturing and we are fierce, we are wicked and we are sublime. The full range is ours. We hold the moon in our bellies and fire in our hearts. We bleed. We give milk. We are the mothers of first words. These words grow. They are our children. They are our stories and our poems.

By allowing ourselves to undress, expose, and embrace the feminine, we commit our vulnerabilities not to fear but to courage—the courage that allows us to write on behalf of the earth, on behalf of ourselves.

—15—
The Boy Who Lived with the Bears

A Mohawk Lesson Story

Joel Monture

For Native Americans, children represent our hope and future. Sadly, the many inroads made by poverty, infant mortality, alcoholism, welfare, poor education, and degradation to our environments by corporations create constant struggles for survival in an altered world. Across the Americas, indigenous people do survive, with many filling roles in "mainstream" culture, from surgeons to professors, artists, attorneys, and a foot in the door of every profession—but not at the cost of tradition.

When I was a graduate student in the Native American Program at Dartmouth College, I experienced an interesting phenomena. I lived with my wife and four-year-old son in graduate housing bungalows. Within our first month at the school our daughter was born. There were few, if any, Native American graduate students—most of our neighbors were young families in either the medical school or the business school, and there was a playground in the middle of the complex. Generally speaking, while the husbands were in classes, the wives came to the playground in the morning, and sat with cups of coffee at picnic tables while the children played on the slide or the swings. My wife would nurse our baby girl, Mary, while our son, Joseph, played in the sand box. Over a period of that first autumn in that community of people who had come from all over the country (representing many religions and economic backgrounds), the tone of conversation among mothers (and fathers who took breaks) was how "stressed out" they were with their "kids." They all seemed to want to be away from their children—as in, "I want my own life back!" This was confusing to us. We were celebrating all the life that had come before us, and was now rushing through the veins of our children into the future of our culture. We thought our children would make us immortal! And they do . . . because there is a grandmother's little smile in the baby girl!

The other Native students who came to the house would ignore their counseling and become distracted holding the baby. They would take a "time out" to give Joseph a ride on the back of the bike. Often they would admit homesickness, and help with the cooking, my wife being an older sister, even though they were of different tribes. What mattered was that there was some unity of family, caring.

On the other hand, non-Natives were perfectly comfortable to give a cursory notice to our children and then get down to the "bottom line," the business at hand, even if that business was social. The notion seeming to be that the activities of adults were not compatible with the activities of children. Within a Native community, a social evening means that all members are welcome, and this is the way children learn to become members of that society. Part of the problem in Western culture is that adults don't seem to think that children have much to say, but in Native life even an infant . . . with a cock of the eyebrow, or a little gas smile is a message. It's about listening, and who you choose to listen to. Indian people have always listened to children, and we have always listened to elders.

Other cultures seem to listen only to their peers.

If you have time. . . . If you can listen. . . . This is a story that has been told for many generations, and it is about taking care of your children.

There used to be a Mohawk village of longhouses, located beyond the pines beside a river. There was a Boy who lived there, and both of his parents had died. As was the way, the Boy lived with his closest relative, his Uncle, who was a hunter of great reputation.

Now, the Uncle, perhaps because his thoughts were always on the business of hunting, had a twisted mind, and he did not like the Boy. He did not care for him, gave him only scraps of food, and barely clothed him. All the people in the village noticed how badly the Uncle treated the orphan Boy, but they said and did nothing, because it was not the Mohawk way to interfere or pass judgment. It was known that the ability to be of the *good mind* came from within, and that everyone was capable of changing.

Still, the Boy loved his Uncle and wanted nothing more than to please him. So when a day came that the Uncle said, "You, Boy, are coming hunting with me today!" the Boy was overjoyed to set out with his Uncle from the village into the deep forest.

Right away the Boy noticed two strange things that puzzled him. In those days it was known that unusual things happened in the north—the hunters always went east or south or west, but never north, and on this day, watching the sun, the Boy saw that they were traveling north through a land that began to change into crooked trees and jagged rocks, crying winds and grey clouds. The Boy asked his Uncle about this, and the Uncle said only, "You, Boy, shut up!"

The Boy noticed also that his Uncle had left his hunting dog tied to a tree back at the village so he wouldn't follow or help in the hunt, and he asked his Uncle about this too.

His Uncle said, "You, Boy, will be my dog today. Dogs don't talk. Shut up!"

So the Boy obeyed, because he had been taught by his Mother and Father to respect elders, and he wanted his Uncle to love him.

They traveled far far to the north, and the world took on stranger and stranger appearances, crooked and twisted, like his Uncle's mind. It was dark and cold here, and they came to a clearing, a circle of bent trees and sharp stones. On the other side of the clearing there was a black hole in a hillside, and the Uncle said, "There is an animal in that cave, dog! Go in and howl, chase out the animal so I can kill it!"

The Boy looked at the dark hole, then he looked at his Uncle. He was afraid of both, but he remembered his lessons, and obeyed, crawling into the mouth of the cave.

"Go down, and howl!" ordered the Uncle, and the Boy, on hands and knees, began to bark out like a dog and descend into the hole.

It was black. He could see nothing. And looking back he saw the circle of light from the opening grow smaller and smaller the deeper he traveled. He continued to bark and crawl, feeling very small, until he reached the bottom of the hole. Feeling around, he found that there was no animal, not even a mouse, and when he turned around the small circle of light suddenly disappeared. He crawled faster and faster toward the mouth of the cave and when he reached the top of the hole he felt a big rock lodged in place, leaving him trapped alone in darkness. Now he understood that his Uncle meant to throw him away, leaving him buried in the earth.

He began to cry, swaying and hugging his knees, and feeling all alone in the world. He thought about his Mother, and he wished she was there to comfort him. Then he remembered his Mother teaching him a little song that he should sing when he felt alone and needed a friend. So he began to sing softly, almost a hum . . . "Way yanna way yanna way yanna ho, way hey yo-oh-oh, way hey yo. . . ."

His small voice seemed at first to come back to him like an echo in that pitch black tomb, but as he sang louder he realized that someone else was singing on the outside, beyond the boulder that imprisoned him. So he sang louder and louder, and with each chorus the voice beyond the rock sang louder! Faster and faster they both sang, and then in a great rumbling roar and rush of air the stone was pulled aside and the Boy was blinded with light! He covered his eyes, feeling pairs of hands grab him and drag him out into the sweet air. He breathed deeply and opened his eyes, trying to adjust to the light. Slowly, gradually he began to make out the forms of many people filling the clearing—a huge crowd of people, fat and thin, tall and short, people of all sizes and types, and when his eyes became accustomed to the light he saw that all the trees were straight and full, enwrapped with flowering vines that gave off such sweet scents. The grass was thick and green, the moss rich on the soft round stones.

Suddenly one of the people scurried up to him sitting on the bank before the cave. She chattered quickly!

"I am Grandmother Woodchuck. . . . What are you doing here?"

The Boy's face was still wet with tears, and he sadly said, "My Uncle threw me away. . . . I am nobody now, for I have no family. No one loves me!"

Grandmother Woodchuck said, "We heard your song! You are not alone . . . you have us, all of us! All you have to do is choose your family!"

Suddenly, little Mole raced about and ran all over Boy's knees, tickling him and making him laugh. "Come with me," piped little Mole, "I will raise you, and we will have a wonderful life together, digging tunnels and eating the most delicious worms! You will love to be a mole!"

The Boy thought about it, and remembered his Father's teachings to always to grateful and show respect. "Oh, Mole, how can I thank you?" (The thought of eating worms was not very pleasant to the Boy.) "But I don't have claws to dig, and I am used to living above the earth."

Little Mole scurried up his shoulder and whispered in her ear, "You are a good Boy!" Mole raced back into the crowd.

Beaver waddled up. "You would love to be a beaver! Come with me and we will swim and splash, and we will eat the most succulent shoots and gnaw down trees. It will be a wonderful life!"

Boy again remembered his respect and said, "How can I thank you? But I can't hold my breath long enough to be a beaver, and my teeth are not as sharp as yours."

Beaver gave him a friendly slap of her tail, saying, "You are a good Boy!"

Then Squirrel scampered up, saying, "You would love to be a Squirrel! All day long we would race through the trees and chatter at the bluejays! We would eat the most delicious nuts, and live high in the branches of the tallest trees!"

Boy thought a second, and said, "How can I thank you enough? But I don't have a long fluffy tail to help me keep balance. . . . I am afraid I would fall!"

Squirrel flickered her tail and made a chattering laugh, "You are a good Boy!"

One by one all the creatures of the forest told the Boy about their lives, but none seemed quite right to him, until Bear came up to him.

She said, "You would love to be a Bear. All day long we amble through the forest, and when we get tired we lie in the sun and rest. We eat the most delicious honey the bees gather for us, and sweet fish from the rivers. And you would never be lonely, because my two cubs will be your brother and your sister!"

The Boy didn't have to think for long. "I would love to be a Bear!" he said. "How can I thank you enough?"

"Come, Son," said Mother Bear, and with that all the creatures of the forest slipped away to be themselves, and the Boy ran off with his Mother, his brother, and his sister.

It was exactly as Mother Bear had said. They traveled in the forest, pawing berries from the bushes, catching fish, lying in the sun, playing and tumbling with his cub brother and sister. But like children who play and wrestle, the Boy was scratched by the claws of the cubs, and everytime he was scratched black fur would grow, until finally he was covered with dark hair. His fingernails grew, and he was able to paw apart rotten stumps and climb trees. In time he had become a Bear. The bees didn't bother him when he muzzled their hives, and he splashed up fat trout from the streams. Life was sweet, and he was loved by his Mother. There were no worries in life until one day Mother Bear stood up on her legs and listened, sniffing the air.

So did the cubs, and when Boy stood up it felt strange.

Mother Bear dropped down, "That is just a hunter we call *Bumps into trees*. He is nothing to worry about."

When Boy listened he heard a hunter crashing in the forest, cursing his luck.

Life continued with bright days and Bears laughing, until one day Mother Bear stood up again, but she quickly dropped down. "That's just a hunter we call *Falls in the lake*."

When Boy listened he heard someone splashing and flailing in the water beyond the ridge.

After a time, Boy came to know all the sounds and smells of the hunters, and they would just turn away, ambling through the forest quietly until it was safe to play or forage in peace.

One day Mother Bear stood up in alarm, listening intently.

Boy listened too, and he heard the sound of footsteps . . . not quite two feet moving through the forest, and not quite four feet shuffling the leaves.

Suddenly, Mother Bear's eyes opened wide and for the first time Boy saw a look of fear on her face as her ears dropped back. "That is Two Feet and Four Feet, Very Bad! *Run!*"

All four Bears began to climb the ridges, running back, trying to double back off to the side to confuse the hunter, but always when they paused to listen they heard the hunter and his dog close behind, cutting across the trail towards them! The dog was braying and growly, and the hunter was urging the dog on after the bears. They ran and ran, until finally Mother bear dived into a wide hollow log laying on the ground among the ferns and pines. Inside they all waited and lay still, silent, keeping their breath shallow. Outside everything was quiet as they waited.

It did not take long. Inside the log the Bears sniffed. They sniffed again the bad smell now coming into the log. It had been a long time since the Boy had smelled this, but it slowly came back to him as the smoke filled their dark space. Mother Bear began to snort, and the cub began to sneeze, tears filling their eyes from the hunter's fire. Boy turned towards the opening of the log, wondering what could be done to save his family. He thought and thought, and finally he remembered something . . . he remembered from far back that he had once been a human being, and he remembered how to speak.

He called out, "Please, don't harm us!!"

Slowly he crawled out of the log and stood up like a human.

In front of him was a hunter, and the hunter's dog stood on braced legs snarling at him. The hunter had his bow drawn back far and the arrow pointed at the Boy's heart. The boy saw that it was his Uncle who had thrown him away.

He said, "Uncle, since you abandoned me, I have become a bear, and our family is filled with love. Please, let us go away now."

The Uncle's face was full of disbelief to see this black bear speaking before him, and he slowly lowered his bow and cuffed back the dog. The Uncle reached out a hand and touched the bear, and when he did, all the hair fell off the Boy's body to reveal his nephew.

Now the Uncle began to weep. The tears rolled down his cheeks, and he dropped his bow. He trembled like a small tree in the wind, and threw his arms around the Boy.

"My mind was twisted, and I did a bad thing," he cried, "And when I realized it I came back for you! But the stone was rolled away, and the earth was covered with animal tracks, and I knew the animals had dragged you away and eaten you! How can you ever forgive me?"

The Boy, standing naked, accepted his Uncle's robe and covered his shoulders. He kicked away the fire and called out his family.

Slowly, his Mother, his Brother, and his Sister emerged and they walked past Uncle and his dog. The Boy watched as they climbed a ridge, and when they briefly stopped and looked back, their hearts and minds were filled with the sadness of parting, the memories of experience that would never happen again, and the knowledge that the Boy's time to go back to his own animals had come. Boy watched as they disappeared into the underbrush.

The Uncle and the Boy sat down to share corn, burn tobacco, and offer prayers of thanks for his return. When they returned to the village there was a great welcome for the Boy, and the Uncle brought a message. . . . "It is as important to love and care for your children as much as Mother Bear loves hers."

Now, some people believe that the reason Bear travels across so much territory is because she is always seeking her missing child.

Let us be as loving as Mother Bear. . . . Let us hold our children close. . . . Let us continue to give. *Nia: wen Kowa*. Thank you.

Part III
—Living Every Day—

&

Love in Action

Thich Nhat Hanh

On Simplicity

In 1951, I went with a few brother monks to a remote mountain in the Dai Lao region of Vietnam to build a meditation center. We asked some native mountain people for their help, and two Montagnards from the Jarai tribe joined us in clearing the forest, cutting trees into lumber, and gathering other materials for construction. They were hard workers, and we were grateful for their assistance. But after working with us only three days, they stopped coming. Without their help, we had many difficulties, as we were not familiar with the ways of the forest. So we walked to their village and asked what had happened. They said, "Why should we return so soon? You already paid us enough to live for a month! We will come again when we run out of rice." At the time, it was a common practice to underpay the Montagnards, to avoid just this kind of thing. We had paid them properly, and, surely enough, they stopped coming.

Many people criticized the Montagnards for this ethic. They said that this laziness could only lead to trouble, and they listed four reasons to support their claim: (1) The Montagnards would be happier and more comfortable if they would work harder. (2) They would earn more money, which they could save for difficult periods. (3) The Montagnards should work harder in order to help others. (4) If they would work harder, they would have the means to defend themselves from invasions and the exploitation of others. There may be some validity to each of these points, but if we look closely at the lives of the Montagnards, we will come to understand them, and ourselves, better.

1. The Montagnards would be happier and more comfortable if they would work harder.
The Montagnards lived simply. They did not store much food at all. They had no bank accounts. But they were much more serene and at peace with themselves, nature, and other people, than almost anyone in the world. I am not suggesting that we all return to primitive lifestyles, but it is important

that we see and appreciate the wisdom contained in a lifestyle like this, a wisdom that those of us immersed in modernization and economic growth have lost.

How much stuff do we need to be happy and comfortable? Happiness and comfort vary according to taste. Some people think they need three or four houses—one on the Riviera, one in New York, one in Tokyo, and perhaps one in Fiji. Others find that a two or three-room hut is quite enough. In fact, if you own a dozen luxurious houses, you may rarely have time to enjoy them. Even when you have the time, you may not know how to sit peacefully in one place. Always seeking distraction—going to restaurants, the theater, or dinner parties, or taking vacations that exhaust you even more, you can't stand being alone and facing yourself directly.

In former times, people spent hours drinking one cup of tea with dear friends. A cup of tea does not cost much, but today, we go to a cafe and take less than five minutes to drink our tea or coffee, and even during that short time, we are mostly thinking and talking about other things, and we never even notice our tea. We who own just one house barely have the time to live in it. We leave home early in the morning after a quick breakfast and go off to work, spending an hour in the car or the train and the rest of the day in the office. Then we return home exhausted, eat dinner, watch TV, and collapse so we can get up early for work. Is this "progress"?

The Montagnards were quite content to live in simple bamboo and palm-leaf huts and wash their clothes by hand. They refused to be slaves to economic pressures. Content with just a few possessions, they rarely needed to spend their time or money seeing doctors or psychotherapists for stress-related ailments.

2. They would earn more money, which they could save for difficult periods.
How much do we need to save? We do not save air, because we trust that it will be available to us when we need it. Why must we stockpile food, money, or other things for our own private use, while so many others are hungry?

People who accumulate a house, a car, a position, and so forth, identify themselves with what they own, and they think that if they lose their house, their car, or their position, they would not be themselves. To me, they are already lost. By accumulating and saving, they have constructed a false self, and in the process they have forgotten their truest and deepest self. Psychotherapists can try to help, but the cause of this illness is in their way of life. One way to help such a person would be to place him in an "underdeveloped" country where he could grow his own food and make his own clothes. Sharing the fate and simple life of peasants might help him heal quickly.

We have enough resources and know-how to assure every human being of adequate shelter and food every day. If we don't help others live, we ourselves are not going to be able to live either. We are all in the same boat— the planet earth. Why not put our efforts into trying to help each other and save our boat instead of accumulating savings only for ourselves and our own children?

3. *The Montagnards should work harder in order to help others.*

Of course, the Montagnards could have spent more time working in order to send aid to people who were starving in other parts of the world. If they did not do so, it was because they didn't know much about the existence of other nations. They certainly did help their own tribal members whenever they got sick or when a crop was destroyed by some natural disaster. But let us reflect for a moment on what the Montagnard people did not do.

They did not harm or exploit others. They grew their own food and exchanged some of their products with other people. They did not do violence to nature. They cut only enough wood to build their houses. They cleared only enough land to plant their crops. Because of their simple lifestyle, they did not overconsume natural resources. They did not pollute the air, water, or soil. They used very little fuel and no electricity. They did not own private cars, dishwashers, or electric razors. The way they lived enabled natural resources to continually renew themselves. A lifestyle like theirs demonstrates that a future for humankind is possible, and this is the most helpful thing anyone can do to help others.

4. *If they would work harder, they would have the means to defend themselves from invasion and the exploitation of others.*

It is true that the Montagnards were exploited by others and were often victims of social injustice. They lived in remote mountain areas. If others settled nearby, they risked losing their land due to a lack of means with which to defend themselves.

People said that if the rest of us in Vietnam worked as little as they did, our country would never be able to resist foreign intervention and exploitation. It seems clear that the Montagnards and others like them had to do something more. But what? If the Montagnards would have moved down to the more populated areas, they would have seen men and women working extremely hard and getting poorer. They would have seen how expensive food, lodging, electricity, water, clothing, and transportation were. Their civilized countrymen were working all day long and could barely pay for the most basic items they consumed. The Montagnards in the forest did not need to spend any money. If they would have lived and worked in the cities, how would that have helped Vietnam resist foreign intervention? All they would have learned is that in the so-called developed nations, resources are used to make bombs and other elaborate weapons, while many citizens live in misery. The Montagnards might well need nuclear weapons to resist foreign intervention if they were to catch up with their more "developed" brothers and sisters. Will social injustice ever be abolished before all people wake up and realize that unless we let others live, we ourselves will not be able to live?

Economic growth may be necessary for the welfare of people, but the present rate of economic growth is destroying humanity and nature. Injustice is rampant. We humans are part of nature, and doing harm to nature only harms us. It is not just the poor and oppressed who are victims of environmental damage. The affluent are just as much victims of pollution and the exploitation of resources. We must look at the whole picture and ask, "Does our way of life harm nature? Does our way of life harm our fellow humans? Do we live at the expense of others, at the expense of the present, and at the expense of the future?" If we answer truthfully, we will know how to orient our lives and our actions. We have much to learn from the Montagnards and others like them. We must learn to live in a way that makes a future possible.

The Human Family

Although human beings are a part of nature, we single ourselves out and classify other animals and living beings as "nature," while acting as if we were somehow separate from it. Then we ask, "How should we deal with nature?" We should deal with nature the way we should deal with ourselves! Nonviolently. We should not harm ourselves, and we should not harm nature. To harm nature is to harm ourselves, and vice versa. If we knew how to deal with ourselves and our fellow human beings, we would know how to deal with nature. Human beings and nature are inseparable. By not caring properly for either, we harm both.

We can only be happy when we accept ourselves as we are. We must first be aware of all the elements within us, and then we must bring them into harmony. Our physical and mental well-being are the result of understanding what is going on in ourselves. This understanding helps us respect nature in ourselves and also helps us bring about healing.

If we harm another human being, we harm ourselves. To accumulate wealth and own excessive portions of the world's natural resources is to deprive our fellow humans of the chance to live. To participate in oppressive and unjust social systems is to widen the gap between rich and poor and thereby aggravate the situation of social injustice. Yet we tolerate excess, injustice, and war, while remaining unaware that the human race as a family is suffering. While some members of the human family are suffering and starving, for us to enjoy false security and wealth is a sign of insanity.

The fate of each individual is inextricably linked to the fate of the whole human race. We must let others live if we ourselves want to live. The only alternative to coexistence is co-nonexistence. A civilization in which we kill and exploit others for our own aggrandizement is sick. For us to have a healthy civilization, everyone must be born with an equal right to education, work, food, shelter, world citizenship, and the ability to circulate freely and settle on any part of the earth. Political and economic systems that deny one person these rights harm the whole human family. We must begin by becoming aware of what is happening to every member of the human family if we want to repair the damages already done.

To bring about peace, we must work for harmonious coexistence. If we continue to shut ourselves off from the rest of the world, imprisoning ourselves in our narrow concerns and immediate problems, we are not likely to make peace or to survive. It is difficult for one individual to preserve harmony among the elements within himself, and it is even more difficult to preserve harmony among the members of the human family. We have to understand the human race to bring it into harmony. Cruelty and disruption destroy the harmony of the family. We need legislation that keeps us from doing violence to ourselves or nature, and prevents us from being disruptive and cruel.

We have created a system that we cannot control. This system imposes itself on us, and we have become its slaves. Most of us, in order to have a house, a car, a refrigerator, a TV, and so on, must sacrifice our time and our lives in exchange. We are constantly under the pressure of time. In former times, we could afford three hours for one cup of tea, enjoying the company of our friends in a serene and spiritual atmosphere. We could organize a party to celebrate the blossoming of one orchid in our garden. But today we can no longer afford these things. We say that time is money. We have created a society in which the rich become richer and the poor become poorer, and in which we are so caught up in our own immediate problems that we cannot afford to be aware of what is going on with the rest of the human family. We see images on TV, but we do not really understand our Third World brothers and sisters.

The individual and all of humanity are both a part of nature and should be able to live in harmony with nature. Nature can be cruel and disruptive and therefore, at times, needs to be controlled. To control is not to dominate or oppress but to harmonize and equilibrate. We must be deep friends with nature in order to control certain aspects of it. This requires a full under-standing of nature. Typhoons, tornadoes, droughts, floods, volcanic erup-tions, and proliferations of harmful insects all constitute danger and destruction to life. Although parts of nature, these things disrupt the har-mony of nature. We should be able to prevent to a large degree the destruc-tion that natural disasters cause, but we must do it in a way that preserves life and encourages harmony.

The excessive use of pesticides that kill all kinds of insects and upset the ecological balance is an example of our lack of wisdom in trying to control nature.

The harmony and equilibrium in the individual, society, and nature are being destroyed. Individuals are sick, society is sick, and nature is sick. We must reestablish harmony and equilibrium, but how? Where can we begin the work of healing? Would we begin with the individual, society, or the environment? We must work in all three domains. People of different disci-plines tend to stress their particular areas. For example, politicians consider an effective rearrangement of society necessary for the salvation of humans and nature, and therefore urge that everyone engage in the struggle to change political systems.

We Buddhist monks are like psychotherapists in that we tend to look at the problem from the viewpoint of mental health. Meditation aims at creating harmony and equilibrium in the life of the individual. Buddhist meditation uses the breath as a tool to calm and harmonize the whole human being. As in any therapeutic practice, the patient is placed in an environment that favors the restoration of harmony. Usually psycho-therapists spend their time observing and then advising their patients. I know of some, however, who, like monks, observe themselves first, recog-nizing the need to free their own selves from the fears, anxieties, and despair that exist in each of us. Many therapists seem to think that they themselves have no mental problems, but the monk recognizes in himself the susceptibility to fears and anxieties, and to the mental illness that is caused by the inhuman social and economic systems that prevail in today's world.

Buddhists believe that the reality of the individual, society, and nature's integral being will reveal itself to us as we recover, gradually ceas-ing to be possessed by anxiety, fear, and the dispersion of mind. Among the three—individual, society, and nature—it is the individual who begins to effect change. But in order to effect change, he or she must have personally recovered, must be whole. Since this requires an environment favorable to healing, he or she must seek the kind of lifestyle that is free from destructiveness. Efforts to change the environment and to change the individual are both necessary, but it is difficult to change the environment if individuals are not in a state of equilibrium. From the mental health point of view, efforts for us to recover our humanness should be given priority.

Restoring mental health does not mean simply helping individuals adjust to the modern world of rapid economic growth. The world is sick, and adapting to an unwell environment will not bring real health. Many people who seek the help of a psychotherapist are really victims of modern life, which separates human beings from the rest of nature. One way to help such a person may be to move him or her to a rural area where he can cultivate the land, grow his own food, wash his clothes in a clear river, and live simply, sharing the same life as millions of peasants around the world. For psychotherapy to be effective, we need environmental change, and psychotherapists must participate in efforts to change the environment. But that is only half their task. The other half is to help individuals be themselves, not by helping them adapt to an ill environment, but by providing them with the strength to change it. To tranquilize them is not the way. The explosion of bombs, the burning of napalm, the violent deaths of relatives and neighbors, the pressures of time, noise, and pollution, the lonely crowds have all been created by the disruptive course of our economic growth. They are all sources of mental illness, and they must end. Anything we can do to bring them to an end is preventive medicine. Political activities are not the only means to this end.

While helping their particular patients, psychotherapists must, at the same time, recognize their responsibility to the whole human family. Their work must also prevent others from becoming ill. They are challenged to safeguard their own humanness. Like others, psychotherapists and monks need to observe first themselves and their own ways of life. If they do, I believe they will seek ways to disengage themselves from the present economic systems in order to help reestablish harmony and balance in life. Monks and psychotherapists are human beings. We cannot escape mental illness if we do not apply our disciplines to ourselves. Caught in forgetfulness and acquiescence to the status quo, we will gradually become victims of fear, anxiety, and egotism of all kinds. But if psychotherapists and monks, through mutual sharing, help each other apply our disciplines to our own lives, we will rediscover the harmony in ourselves and thereby help the whole human family.

A tree reveals itself to an artist when he or she can establish a genuine relationship with it. If a human is not a real human being, he may look at his fellow humans and not see them; he may look at a tree and not see it. Many of us cannot see things because we are not wholly ourselves. When we are wholly ourselves, we can see how one person by living fully demonstrates to all of us that life is possible, that a future is possible. But the question, "Is a future possible?" is meaningless without seeing the millions of our fellow humans who suffer, live, and die around us. Only when we really see them will we be able to see ourselves and see nature.

The Sun My Heart

When I first left Vietnam, I had a dream in which I was a young boy, smiling and at ease, in my own land, surrounded by my own people, in a time of peace. There was a beautiful hillside, lush with trees and flowers, and on it was a little house. But each time I approached the hillside, obstacles prevented me from climbing it, and then I woke up.

The dream recurred many times. I continued to do my work and to practice mindfulness, trying to be in touch with the beautiful trees, people, flowers, and sunshine that surrounded me in Europe and North America. I looked deeply at these things, and I played under the trees with the children exactly as I had in Vietnam. After a year, the dream stopped. Seeds of acceptance and joy had been planted in me, and I began to look at Europe, America, and other countries in Asia as also my home. I realized that my home is the earth. Whenever I felt homesick for Vietnam, I went outside into a backyard or a park, and found a place to practice breathing, walking, and smiling among the trees.

But some cities had very few trees, even then. I can imagine someday soon a city with no trees in it at all. Imagine a city that has only one tree left. People there are mentally disturbed, because they are so alienated from nature. Then one doctor in the city sees why people are getting sick, and he offers each person who comes to him the prescription: "You are sick because you are cut off from Mother Nature. Every morning, take a bus, go to the tree in the center of the city, and hug it for fifteen minutes. Look at the beautiful green tree and smell its fragrant bark."

After three months of practicing this, the patient will feel much better. But because many people suffer from the same malady and the doctor always gives the same prescription, after a short time, the line of people waiting their turn to embrace the tree gets to be very long, more than a mile, and people begin to get impatient. Fifteen minutes is now too long for each person to hug the tree, so the city council legislates a five-minute maximum. Then they have to shorten it to one minute, and then only a few seconds. Finally there is no remedy at all for the sickness.

If we are not mindful, we might be in that situation soon. We have to remember that our body is not limited to what lies within the boundary of our skin. Our body is much more immense. We know that if our heart stops beating, the flow of our life will stop, but we do not take the time to notice the many things outside of our bodies that are equally essential for our survival. If the ozone layer around our earth were to disappear for even an instant, we would die. If the sun were to stop shining, the flow of our life would stop. The sun is our second heart, our heart outside of our body. It gives all life on earth the warmth necessary for existence. Plants live thanks to the sun. Their leaves absorb the sun's energy, along with carbon dioxide from the air, to produce food for the tree, the flower, the plankton. And thanks to plants, we and other animals can live. All of us—people, animals, plants, and minerals—"consume" the sun, directly and indirectly. We cannot begin to describe all the effects of the sun, that great heart outside of our body.

When we look at green vegetables, we should know that it is the sun that is green and not just the vegetables. The green color in the leaves of the vegetables is due to the presence of the sun. Without the sun, no living being could survive. Without sun, water, air, and soil, there would be no vegetables. The vegetables are the coming-together of many conditions near and far.

There is no phenomenon in the universe that does not intimately concern us, from a pebble resting at the bottom of the ocean, to the movement of a galaxy millions of light years away. Walt Whitman said, "I believe a blade of grass is no less than the journey-work of the stars. . . . " These words are not philosophy. They come from the depths of his soul. He also said, "I am large, I contain multitudes."

This might be called a meditation on "interbeing endlessly interwoven." All phenomena are interdependent. When we think of a speck of dust, a flower, or a human being, our thinking cannot break loose from the idea of unity, of one, of calculation. We see a line drawn between one and many, one and not one. But if we truly realize the interdependent nature of the dust, the flower, and the human being, we see that unity cannot exist without diversity. Unity and diversity interpenetrate each other freely. Unity is diversity, and diversity is unity. This is the principle of interbeing.

If you are a mountain climber or someone who enjoys the countryside or the forest, you know that forests are our lungs outside of our bodies. Yet we have been acting in a way that has allowed millions of square miles of land to be deforested, and we have also destroyed the air, the rivers, and parts of the ozone layer. We are imprisoned in our small selves, thinking only of some comfortable conditions for this small self, while we destroy our large self. If we want to change the situation, we must begin by being our true selves. To be our true selves means we have to *be* the forest, the river, and the ozone layer. If we visualize ourselves as the forest, we will experience the hopes and fears of the trees. If we don't do this, the forests will die, and we will lose our chance for peace. When we understand that we inter-are with the trees, we will know that it is up to us to make an effort to keep the trees alive. In the last twenty years, our automobiles and factories have created acid rain that has destroyed so many trees. Because we inter-are with the trees, we know that if they do not live, we too will disappear very soon.

We humans think we are smart, but an orchid, for example, knows how to produce noble, symmetrical flowers, and a snail knows how to make a beautiful, well-proportioned shell. Compared with their knowledge, ours is not worth much at all. We should bow deeply before the orchid and the snail and join our palms reverently before the monarch butterfly and the magnolia tree. The feeling of respect for all species will help us recognize the noblest nature in ourselves.

An oak tree is an oak tree. That is all an oak tree needs to do. If an oak tree is less than an oak tree, we will all be in trouble. In our former lives, we were rocks, clouds, and trees. We have also been an oak tree. This is not just Buddhist; it is scientific. We humans are a young species. We were plants, we were trees, and now we have become humans. We have to remember our past existences and be humble. We can learn a lot from an oak tree.

All life is impermanent. We are all children of the earth, and, at some time, she will take us back to herself again. We are continually arising from Mother Earth, being nurtured by her, and then returning to her. Like us, plants are born, live for a period of time, and then return to the earth. When they decompose, they fertilize our gardens. Living vegetables and decomposing vegetables are part of the same reality. Without one, the other cannot be. After six months, compost becomes fresh vegetables again. Plants and the earth rely on each other. Whether the earth is fresh, beautiful, and green, or arid and parched depends on the plants.

It also depends on us. Our way of walking on the earth has a great influence on animals and plants. We have killed so many animals and plants and destroyed their environments. Many are now extinct. In turn, our environment is now harming us. We are like sleepwalkers, not knowing what we are doing or where we are heading. Whether we can wake up or not depends on whether we can walk mindfully on our Mother Earth. The future of all life, including our own, depends on our mindful steps.

Birds' songs express joy, beauty, and purity, and evoke in us vitality and love. So many beings in the universe love us unconditionally. The trees, the water, and the air don't ask anything of us; they just love us. Even though we need this kind of love, we continue to destroy them. By destroying the animals, the air, and the trees, we are destroying ourselves. We must learn to practice unconditional love for all.

Our earth, our green beautiful earth is in danger, and all of us know it. Yet we act as if our daily lives have nothing to do with the situation of the world. If the earth were your body, you would be able to feel many areas where she is suffering. Many people are aware of the world's suffering, and their hearts are filled with compassion. They know what needs to be done, and they engage in political, social, and environmental work to try to change things. But after a period of intense involvement, they become discouraged, because they lack the strength needed to sustain a life of action. Real strength is not in power, money, or weapons, but in deep, inner peace.

If we change our daily lives—the way we think, speak, and act—we change the world. The best way to take care of the environment is to take care of the environmentalist.

Many Buddhist teachings help us understand our interconnectedness with our mother, the earth. One of the deepest is the Diamond Sutra, which is written in the form of a dialogue between the Buddha and his senior disciple, Subhuti. It begins with this question by Subhuti: "If daughters and sons of good families wish to give rise to the highest, most fulfilled, awakened mind, what should they rely on and what should they do to master their thinking?" This is the same as asking, "If I want to use my whole being to protect life, what methods and principles should I use?"

The Buddha answers, "We have to do our best to help every living being cross the ocean of suffering. But after all beings have arrived at the shore of liberation, no being at all has been carried to the other shore. If you are still caught up in the idea of a self, a person, a living being, or a life span, you are not an authentic bodhisattva." Self, person, living being, and life span are four notions that prevent us from seeing reality.

Life is one. We do not need to slice it into pieces and call this or that piece a "self." What we call a self is made only of non-self elements. When we look at a flower, for example, we may think that it is different from "non-flower" things. But when we look more deeply, we see that everything in the cosmos is in that flower. Without all of the non-flower elements— sunshine, clouds, earth, minerals, heat, rivers, and consciousness—a flower cannot be. That is why the Buddha teaches that the self does not exist. We have to discard all distinctions between self and non-self. How can anyone work to protect the environment without this insight?

The second notion that prevents us from seeing reality is the notion of a person, a human being. We usually discriminate between humans and non-humans, thinking that we are more important than other species. But since we humans are made of non-human elements, to protect ourselves we have to protect all of the non-human elements. There is no other way. If you think, "God created man in His own image and He created other things for man to use," you are already making the discrimination that man is more important than other things. When we see that humans have no self, we see that to take care of the environment (the non-human elements) is to take care of humanity. The best way to take good care of men and women so that they can be truly healthy and happy is to take care of the environment.

I know ecologists who are not happy in their families. They worked hard to improve the environment, partly to escape family life. If someone is not happy within himself, how can he help the environment? That is why the Buddha teaches that to protect the non-human elements is to protect humans, and to protect humans is to protect non-human elements.

The third notion we have to break through is the notion of a living being. We think that we living beings are different from inanimate objects, but according to the principle of interbeing, living beings are comprised of non-living-being elements. When we look into ourselves, we see minerals and all other non-living-being elements. Why discriminate against what we call inanimate? To protect living beings, we must protect the stones, the soil, and the oceans. Before the atomic bomb was dropped on Hiroshima, there were many beautiful stone benches in the parks. As the Japanese were rebuilding their city, they discovered that these stones were dead, so they carried them away and buried them. Then they brought in live stones. Do not think these things are not alive. Atoms are always moving. Electrons move at nearly the speed of light. According to the teaching of Buddhism, these atoms and stones are consciousness itself. That is why discrimination by living beings against non-living beings should be discarded.

The last notion is that of a life span. We think that we have been alive since a certain point in time and that prior to that moment, our life did not exist. This distinction between life and non-life is not correct. Life is made of death, and death is made of life. We have to accept death; it makes life possible. The cells in our body are dying every day, but we never think to organize funerals for them. The death of one cell allows for the birth of another. Life and death are two aspects of the same reality. We must learn to die peacefully so that others may live. This deep meditation brings forth non-fear, non-anger, and non-despair, the strengths we need for our work. With non-fear, even when we see that a problem is huge, we will not burn out. We will know how to make small, steady steps. If those who work to protect the environment contemplate these four notions, they will know how to be and how to act.

In another Buddhist text, the Avatamsaka (Adorning the Buddha with Flowers) Sutra, the Buddha further elaborates his insights concerning our "interpenetration" with out environment. Please meditate with me on the "Ten Penetrations":

The first is, "All worlds penetrate a single pore. A single pore penetrates all worlds." Look deeply at a flower. It may be tiny, but the sun, the clouds, and everything else in the cosmos penetrates it. Nuclear physicists say very much the same thing: one electron is made by all electrons; one electron is in all electrons.

The second penetration is, "All living beings penetrate one body. One body penetrates all living beings." When you kill a living being, you kill yourself and everyone else as well.

The third is, "Infinite time penetrates one second. One second penetrates infinite time." A *ksana* is the shortest period of time, actually much shorter than a second.

The fourth penetration is, "All Buddhist teachings penetrate one teaching. One teaching penetrates all Buddhist teaching." As a young monk, I had the opportunity to learn that Buddhism is made of non-Buddhist elements. So, whenever I study Christianity or Judaism, I find the Buddhist elements in them, and vice versa. I always respect non-Buddhist teachings. All Buddhist teachings penetrate one teaching, and one teaching penetrates all Buddhist teachings. We are free.

The fifth penetration is, "Innumerable spheres enter one sphere. One sphere enters innumerable spheres." A sphere is a geographical space. Innumerable spheres penetrate into one particular area, and one particular area enters into innumerable spheres. It means that when you destroy one area, you destroy every area. When you save one area, you save all areas. A student asked me, "Thây, there are so many urgent problems, what should I do?" I said, "Take one thing and do it very deeply and carefully, and you will be doing everything at the same time."

The sixth penetration is, "All sense organs penetrate one organ. One organ penetrates all sense organs"—eye, ear, nose, tongue, body, and mind. To take care of one means to take care of many. To take care of your eyes means to take care of the eyes of innumerable living beings.

The seventh penetration is, "All sense organs penetrate non-sense organs. Non-sense organs penetrate all sense organs." Not only do non-sense organs penetrate sense organs, they also penetrate non-sense organs. There is no discrimination. Sense organs are made of non-sense-organ elements. That is why they penetrate non-sense organs. This helps us remember the teaching of the Diamond Sutra.

The eighth penetration is, "One perception penetrates all perceptions. All perceptions penetrate one perception." If your perception is not accurate, it will influence all other perceptions in yourself and others. Suppose a bus driver has an incorrect perception. We know what may happen. One perception penetrates all perceptions.

The ninth penetration is, "Every sound penetrates one sound. One sound penetrates every sound." This is a very deep teaching. If we understand one sound or one word, we can understand all.

The tenth penetration is, "All times penetrate one time. One time penetrates all times—past, present, and future. In one second, you can find the past, present, and future." In the past, you can see the present and the future. In the present, you can find the past and future. In the future, you can find the past and present. They "inter-contain" each other. Space contains time, time contains space. In the teaching of interpenetration, one determines the other, the other determines this one. When we realize our nature of interbeing, we will stop blaming and killing, because we know that we inter-are.

Interpenetration is an important teaching, but it still suggests that things outside of one another penetrate into each other. Interbeing is a step forward. We are already inside, so we don't have to enter. In contemporary nuclear physics, people talk about implicit order and explicit order. In the explicit order, things exist outside of each other—the table outside of the flower, the sunshine outside of the cypress tree. In the implicit order, we see that they are inside each other—the sunshine inside the cypress tree. Interbeing is the implicit order. To practice mindfulness and to look deeply into the nature of things is to discover the true nature of interbeing. There we find peace and develop the strength to be in touch with everything. With this understanding, we can easily sustain the work of loving and caring for the earth and for each other for a long time.

—17—
Ecology: Sacred Homemaking

Thomas Moore

Ecology" comes from two rich and fundamental Greek words: *oikos* and *logos*. *Oikos* refers to the house and home and was used for any kind of dwelling, from a simple tent to a sacred temple. *Oikos* also gives us our words *economy* and *economics*, which have to do with the management (*nomos*) of the home. "Ecology" places *oikos* in conjunction with *logos*, and points us in a somewhat different direction.

Logos is one of the great mystery words we find in many languages, such as *dharma*, *tao*, and *esse*, which we perhaps too plainly translate as "being, law, or nature," for they point to the unfathomable mystery of our own existence and of the world and its things. *Logos* has meant "words, stories, logic, and nature," but it also suggests, at its most profound level, the mysterious essence, or the "quintessence" beyond all materialistic considerations, of a thing. I would define *ecology*, then, simply as "the mystery of home," and suggest that fundamentally it has to do with the soul's constant longing for and establishment of a deep sense of home.

This shift in attention from environment to home takes us out of the realm of nature considered materialistically as a realm of things needing our protection, to a more philosophical consideration of our desire for home and for all the ways in which we can experience it. Home is not merely a physical place; it's an experience that, for all its manifestations in the world, is felt deep in the heart.

It's not uncommon to live in a house that is not a home, just as it's possible to find some satisfaction of the desire for home in something that isn't a house. Some people feel very much at home deep in the woods or high on a mountain trail; others feel more home on a busy city street. Some, of course, find home in both places. Occasionally I sense signs of home in the houses of friends and relatives in ways that I don't find in my own house, and I find myself constantly working on my house—rebuilding, painting, decorating, buying furniture—in an attempt to enrich my experience of home.

The soul has a deep and demanding longing for a heartfelt sense of being at home. It may take months and years to find a house that even begins to satisfy its subtle requirements, and once that house is found, it may not be easy to leave. As a therapist, over the years I've heard countless dreams set in a house that was lovingly described in great detail; apparently, even after we have left a good home it may continue to live with us and in us. People also tell stories full of feeling of the houses in which they grew up or where they once lived, underscoring the lasting importance of these places.

What is this home that is so important to the heart, memory, and fantasy? It's impossible, and not desirable anyway, to explain this mystery, but we can imagine some of its elements. A home provides a sense of containment, belonging, and place—qualities that offer security and identity. Home is an alternative to the heroic journey of being on a quest in work, in the search for intellectual or spiritual knowledge, or in some other area of life. It provides a base and a sense of being settled, at ease, and protected. The quest theme, which can occupy much of our attention, may keep us on the move, while home offers us strong attachment. Both themes may be important in life, but at least in our culture it appears that we need to give special attention to the soul's longing for home, since we already give so much of ourselves to our many projects. Excessive attention to our quests instead of our homes may be responsible in part for our neglect of an ecological way of life.

Home may be encountered in many areas of life, yet at the same time it's extremely personal and intimate. The home we find in dreams is vividly internal, but we might discover some of that same feeling in our house, in the geographical region around us or in our local town, city, or neighborhood. We could even imagine this home soul extending to larger, natural and political areas—a mountain range, a river valley, a county, state, or nation. Ultimately, this same sensibility could embrace the globe and beyond. And all of this is ecology—the mystery of discovering, creating, and sustaining the experience of home.

Taking care of our houses, keeping us close to the fantasy of home, could be the basis for a far-reaching emotional ecology. The more abstract and distant our view of "the environment," the less soulful is our attachment and our attention. Abstraction of all kinds is often a defense against the challenge and power of the soul, which is largely vernacular—of a particular place. Taking care of our homes and neighborhoods is vernacular, saving the planet can be abstract.

We can't discover the world as home abstractly. Classifications, information, theories, and experiments may satisfy the mind, but the soul has very different needs. It works by means of attachment, desire, pleasure, imagination, fantasy, and strong feelings of relatedness. In a culture that values mind over heart in most things, it may take a decided shift in values to move from a mental to an erotic view of ecology—erotic in the sense that our longing, attachment, and intimacy with place is considered more important than abstract ideas and ideals.

I realize that some people can become attached to their own homes and regions in a way that is narcissistic and xenophobic, but the obvious soullessness in that attitude suggests that they are not truly at home, since they have to insist on their houses or locales defensively. Individuality is not achieved by defending oneself against community; on the contrary, it can only be found in the context of community. So, while others are finding their own way "home," we can enjoy ours.

At the beginning of the penultimate chapter of the Odyssey, when Odysseus has returned from his lengthy journeys that had been filled with a longing for home, the servants rush to tell his wife Penelope that he has arrived. "Odysseus is here," they say, "he is at home (*oikos*), though he is late in coming." He was ten symbolic years late, as are we all as we try to cure our absolute homesickness. *The Odyssey*, a wondrous book about mysteries of wandering, longing, family, and marriage, reveals that one of the most profound mysteries in human life is the desire for home and the necessity for each of us of *homecoming*.

Recalling the etymology from *oikos*, we can see that making a home is one of the mysteries at the core of ecology. The end of our maltreatment of nature may come only when we have discovered that this earth, in all its particular places and presences, offers the soul relief from its aching search for a home. The abuse of the environment could be seen as a symptom of our deep-seated and symptomatic homesickness, since we often attack that which we desire but cannot find a way to possess.

Our culture, committed to principles of modernism, believes largely that study and applied experiments are the only trustworthy guides to being in the natural world, but soulful cultures know the importance of rituals, stories, graphic arts, dance, music, mythology, and many other imaginal means, for establishing a vivid feeling of home. As long as our attempts to become more ecologically minded and sensitive rely on modernistic values, we will remain alienated from nature and will express our alienation aggressively and destructively. What is called for is a radical shift in paradigm, a true postmodern way of living that re-evaluates fundamental modernistic assumptions, explores intelligent and sensible alternatives, and focuses more on the heart and the imagination than on the mind with its literalistic applications.

I mention "sensible alternatives" because once we have the intuition that there is something missing in our modernistic way of life, the temptation may arise either to seek out ways of transcending modern methods—ritualism, spiritualism, new ungrounded religions and cults, pseudo-sciences, and the rest—or to take a sentimental approach to nature in which our dark, aggressive engagement with the world is denied and our intelligence jettisoned. It is possible, I believe, to avoid these two extremes, and instead take art and religion seriously and intelligently, allowing them to forge the attachments to the natural world and to the rest of culture that we need. After all, when we define ecology as home, we have to include culture as well as nature among our concerns.

As it stands, both religion and the arts are in a crippled condition and can't effectively perform their roles as gateways to ecology. These days many people of good will are either emotionally obsessed with religion and therefore blinded to intelligence in forming their values, or they are jaded and distrusting in relation to religion. Some are fighting a crusade against secularism, and some are making secularism their religion and philosophy. The arts are in limbo, abandoned to museums that, for all their efforts to be involved in the community, are still at the edge of life, or to galleries and auditoriums for the elite. Artists themselves often can't seem to find a viable philosophy of art amid their many theories. Many are still squeezing out images in the name of self-expression and political statement, neither of which invites much expression of soul.

There is an immediate and powerful relationship between art, religion, and ecology. Ecology is a sensibility, not a political position, and it requires profound education and initiation. Soulful education in ecology involves an evocation of home that may start on the surface as responsibility and the pragmatics of living, but must go deeper in order to satisfy the soul. Deeper levels would include an attachment to place, ways of sustaining memory, and stories about places and objects that give them fantasy and therefore a basis for relationship. Relatedness, whether among people, things, or places, requires story and memory which ultimately lead to value and reverence.

A maturing of ecology also requires a different attitude toward religion than the one currently in fashion. Now we're concerned with belief, tenet, theological explanation, and moral prescriptions. Our focus on the mind of religion fosters competitiveness and intolerance, as well as a serious neglect of concerns of the soul. Whatever one's belief may be, one could expect religion to teach and to foster ways of being in the world that are particularly religious, such as reverence, ritual, remembrance, and honor of the spirits of place. We could learn from religion to praise, celebrate, remember, and re-present ritually the soul of whatever concrete reality gives us the spirit of home. In all of our activities, only religion and art touch the soul, and so everything we do has to be artful and conducive to a sense of the sacred.

II

Shortly before beginning this writing on ecology, I made a trip with my family to Ireland. We spent a few days by the sea in County Clare, and then we stayed for a week in a lovely country-style cottage on the Dingle Peninsula in County Kerry. Our decision to go was intuitive. We don't know where the idea came from, but we took it seriously when it made its appearance, knowing somehow that it was a fantasy worth heeding. Now that the trip is over we can look for reasons.

My great-grandparents came to America from Ireland, and my wife's grandparents are buried in Connamara, County Galway. My daughter, a little red-head of one-and-a-half on this trip, Siobhan, has an Irish name that came to us much the same way as the idea to travel to the "Old Country." In my seminary days, when I was nineteen and twenty, I lived in Ireland for two years and had many experiences that stayed fresh in my memory. So, we have many attachments to Ireland.

We were hoping to find some enchantment. On our first night in Doolin by the sea, we took a walk up a hill just away from the B & B, toward the ocean. The road made a sharp turn and there we found before us a jewel of a castle, high on the hill and overlooking the sea. Except for a small wall and sidehouse, the entire castle consisted of a round tower in pristine condition. We learned later that it was privately owned—the reason for its mint condition and for it being inaccessible to the public. We lamented the fact that such a beauty of a castle from the fifteenth century was excluded from the common enjoyment. One would hope that in an ecological spirit a community would know that certain objects cannot be privately owned but belong to the people as the furnishings of their home.

Nevertheless, the castle was our first encounter with enchantment and breath-taking beauty. Another was the discovery that the Irish live close to animals, and that throughout the land their companionship is never far away. Cows grazed at the fence of our cottage; goats and sheep, unfettered by fences, lined the roads in many of the places where we traveled; and every evening we looked out our window to see a neighbor's horse and colt walk in silhouette against the late setting sun, as thick clouds poured over the mountains behind them presenting an ethereal backdrop. More than once the road was blocked by a convention of animals walking very slowly to an unknown destination, so that we had to drive as one beast among others, the cows staring us down inches from our noses.

We bought a mimeographed pamphlet in a local village that told of certain out-of-the-way sites that could be found by noting the color of barn doors, the size of pasture gates, and the direction of the bends in trees. Following the colorful guide we found a castle not privately owned, on the edge of the sea, looking as though it had been sieged and sacked just yesterday, still guarding the inlet from Vikings. Its clear stream and verdant herb patch were still fresh and brilliant, and gave us vivid images of life on this bountiful piece of land. Following obscure, inefficient, and labyrinthine directions, we came upon St. John's Well, guarded as is traditional by a dark, overshadowing Hawthorne tree. We blessed ourselves and the children with its holy water, no doubt both pagan and Christian—doubly sanctified—and took in the palpable, overwhelming sacredness of the place. "This is what brought us to Ireland," my wife said, as my daughter swizzled her wool-stockinged foot in the well.

As an American, I felt that the Irish enjoyed an attachment to their land and traditions, and conveyed a sense of national community with an emotional intensity that I would be surprised to find in my own country. Their feeling was not patriotism in the political or moralistic sense, but a strong evocation of family. They set me wondering about the relationship between ecology and family spirit. I know that Americans are often proud of their regions and towns, but as a nation we seem more interested in ideals and principles than in genuine affections, and this quality carries over into our humorless ways of being ecological.

Remaining loyal and attached to histories and traditions, to the spirit that sings in a place's music and painting, is part of ecology. In our society we make too much of a break between nature and culture. We want to take the lead and shape nature to our designs, but there is another way: We could allow ourselves to be fashioned by the climate, the geology, and the colors of our particular place. We could maintain our loyalties to the culture that raised us, as well as to others we've adopted, creating a rich tapestry of loyalties and customs.

In modern life, we often think that it's so important to be an individual that we suppress our traditions. Yet, it is our traditions, drawn from the land and sky and from the imagination of people who have lived on that land for eons, that give us soul, and it is by being soulful that we are most truly individual. People who are destructively aggressive and neglectful in relation to nature may be harboring fears that nature is the enemy, that she will swallow us up. Generally there seems to be little trust in nature, both nature as the outside world and as our own inner natures. Ecology requires deep trust in both kinds of nature, inner and outer.

I mention my journey to Ireland not to suggest that this particular land can offer everyone a way toward ecology, but because my own roots and my family's roots are there. Each person has a particular background and loyalties that offer a basis for an ecological sensibility. The mind can be abstract and general in its survey of sacred places, but the soul wants its attachments to family and soil to be respected and taken seriously. The soul is always vernacular, speaking the language and customs of a particular place. Our ecology, too, has to be vernacular if it is to invite soulfulness; it might be helpful, therefore, even as we talk about ecology to keep our own places, homes, and heritages warm in our words and thoughts.

III

Let us return to reflection on the soulful experience of home—the very core of an ecological sensibility. This experience is archetypal. By that I mean that it is not derived only or even primarily from our childhood experiences of home. Many people have unhappy memories of that home anyway. The experience can be found in houses and on certain pieces of land, but home is not literally about a building or a location. Home can be evoked by a song, a smell, the taste of certain food, family stories, and a special kind of companionship and community. The soul always has a need for home, and this desire may never be fully satisfied. Even when we have found our dream house, as we so tellingly call it, there is still more home to be mined from life and nature. It is as though a home is forever being built in the soul out of the many different experiences that evoke it.

When life has been made soulful through the many experiences that touch the heart and stir the imagination, then the distance between inner and outer diminishes. With a sense of home set well in the soul, we are able to find it more readily and feel it more intensely in the world around us. We learn what it takes to evoke a feeling of home in different situations. When I lecture I often place photographs of my family on the lectern in front of me so that in my work I'm not cut off from that important dimension of home. There is an ecology of public speaking, as there is of everything we do. Ecology is not one activity among others, it is a dimension of all things. We could each reflect on the ecology of our work, family, house, region, entertainment, self-expression, sports, and social life, asking how the soul's longing for home is addressed in each of these areas.

I realize as I talk about simple things like caring for our houses and finding home in our daily work that many people are placing their lives on the line to protect animals, forests, and seas. One might well ask if my concerns are not too small in relation to these important responses to our global self-destruction. It's my position that true ecological activism always has to be rooted in the concerns of the soul, and those concerns tend toward the individual, the family, and the locality. We can go abroad with our activism, but we may need to ground our ecology close to home, so that our broader activism does not lose its soul.

Ecology is the mysterious work of providing home for the soul, one that is felt in the very depths of the heart. In the world, that work necessarily includes care of one's own life and soul as well as tending carefully to one's house, family, neighborhood, town, region, country, the planet, and other bodies of the universe. Neglect or abuse of any of these parts of home threatens the soul's need for security and place. Once we have the imagination that sees home in such a profound and far-reaching sense, protection of the environment will follow, for ecology is a state of mind, an attitude, and a posture that begins at the very place you find yourself this minute, and extends to places you will never see in your lifetime. The description of divinity ascribed to the mythological magus Hermes Trismegistus and repeated by Neo-Platonists down the centuries applies to our view of ecology: "God is a sphere whose center is everywhere, and whose circumference is nowhere." The object of our ecological concern is nothing less than that sphere, and yet it is felt as the most intimate enclosure and embrace.

—18—
India's Earth Consciousness

Christopher Key Chapple

The environmental process begins with consciousness. To be aware of one's ecosystem, one must begin with the body, the seat of consciousness.

In ancient India, the person or *puruṣa* was regarded as a reflection of the world: eyes were the sun and moon; breath, the wind; feet, the earth. By looking closely at one's own body, the cosmos itself could be discerned. The relationship between the microphase of one's bodies and the macrophase workings of the universe provides a root metaphor for seeing the world from a wholistic perspective, leading to environmental awareness. By seeing the universe as reflective of and relating to body functions, one sees oneself not as an isolated unit but as part of a greater whole.

The *Rig Veda* speaks of consciousness in terms of two birds sitting in the same tree: one eats sweet berries on one branch, while the other gazes dispassionately from a different branch. This two-fold analysis of reality led to a philosophical system known as Sāmkhya. The active bird, associated with the female gender, comes to be known as *prakṛti*, while the inactive bird is referred to as *puruṣa*, linked with the male gender. *Prakṛti* manifests her activity through three all-pervasive modalities (*guṇas*) that in turn divide into twenty-three concrete realities (*tattvas*). These realities are brought to life through the power of consciousness, for whom the manifestations of *prakṛti* dance alluringly.

These twenty-three begin with operations of the mind: emotion (*buddhi*), ego (*ahaṃkara*), and thought (*manas*). Four sets of five then issue forth from one's mental structures and predispositions: the senses, the body, the subtle elements or capacities, and the gross elements. Smelling, tasting, seeing, touching, and hearing (the senses) attach themselves to walking, grasping, talking, excreting, and sexual capacity (the body). This bundle then goes forward through its subtle powers of perception to make contact with the earth, with water, with fire, with air, with space (the gross elements).

Simply put, according to Sāmkhya, the world begins with structures of consciousness that unreel themselves such that they construe and engage physical reality. Emphasis is placed not on the world as a given, external reality, but as an expressive experience of each individual's propensities and attachments. The world hinges on taste and preference; one sees what one intends to see, and becomes according to one's own predilections.

The world in this system is intimate to one's own sense of being. Without the sense of smell, there could be no crisp scent of autumn leaves, no gentle wafting of the ocean breeze. Without the sense of taste, there could be no soothing drink of water or startling spiciness. Without the eyes and the light that bathes them, these could be no color or form. Without the garment of skin that cloaks our bodies, there could be no caress of human contact. Without the ears, no sound could be heard or sent forth as communication with others. In the most intimate of ways, the world cannot exist without the body in which we dwell.

Reflecting on the power and glory of the body, practitioners of Yoga developed techniques by which one could finely tune the body's capacity for sensory experience and also for directed thought. Through the application of postures that imitate animals and through the controlled use of breath, the human person is guided through Yoga to experience the world from the perspective of other living beings. By holding the lion's pose, one approximates the experience of ferocity and fullness. By standing in the tree posture, one gains the stability and rootedness of the tree. By mastering the breath, one learns the origin and control of thought.

This intimacy of thought with body and of body with the world, can be seen from an ecological angle. This outlook does not allow for a world to exist outside of one's perception of it. It can help us recognize and remember that our very sustenance must not be taken for granted. The food we eat is part of the same elemental structure of which we are composed. Furthermore, we depend on the labor of other humans who produce and transport food for our consumption. These people, in their work, must rely upon soil, sunshine, and rainfall to produce our food. We cannot conceive of anything in the universe that does not rely on and relate to the creative expression of the five elements; nor can we have access to any item without the power of our own sensuality.

When one can see and smell a flower in such a way that the flower becomes a celebration of the creative powers of the universe manifested both internally in human biology and externally in the botanical realm, then a true sensitivity to the way of the earth can be cultivated. An environmental ethic emerges from a sacred attention (_puruṣa_) to the needs of one's body and the earth itself, both of which become manifest through the creative matrix (_prakṛti_).

Buddhists of long ago developed a meditation to cultivate appreciation of one's ephemerality, advocating one to visualize first oneself and then all others as nothing more than rotting flesh giving way to bare bones. Through applying this technique, one was empowered to overcome attachment to one's physicality and the attractiveness of others. Rather than stripping away our corporeality, modes of meditation need to be developed and practiced that can re-enflesh our reality and reveal our intimacy with universal processes. Just as Thomas Berry has advocated turning away from Christianity's obsession with personal sin toward a celebration of creative processes in the universe, so also in the Asian traditions there is now room and a need for applying their profound interiorities in such a way as to heighten awareness of the earth's post-industrial dilemma. Gurani Anjali has stated that Americans need to wake up to their senses in order to see the world; it is only by clearly seeing the world that wisdom can be attained. Through breathing deeply, walking quietly, tasting delicately, hearing keenly, lessons can be gleaned from the world about the world. As our perceptions become clarified, thoughts become still. As our thoughts become still, our tendency to acquisitiveness begins to subside; a contentment arises that can serve as the perfect antidote to the over-consumption that currently chokes our culture. By centering ourselves we turn from the outward-moving immaturity that seeks constant gratification. To center ourselves, we need to return to the power of our own senses and their relationship with the elements that compose both our bodies and the planet itself. As our attachments weaken, our consumption will diminish. The diminishment of consumption in turn is the key to an environmentally sensitive lifestyle.

For the early Buddhists, and within certain forms of Yoga, the observance of the elements (*kasinas*) held a key place within the training and cultivation of consciousness. Meditation on the earth, on water, on fire, on air, and on space served several functions. It reminded the practitioner of the commonality of elements: all persons and all things are composed of these essential components. A particular configuration, when seen in its glorious individuality, might seem appealing, might become an object of desire or lust, whether it be an attractive house or car or even body. But when seen from the perspective of its sharing with all other entities in the universe its basic composition, it might be viewed with a bit more reserve, with a bit more care. When a perception of common origin colors one's perspective on, or perception of, things in the world, then one is less likely to give oneself over to infatuation or attachment.

In the traditional Buddhist and Yogic way of meditating or concentrating on the elements, the practitioner fashions in succession a series of supports for honing in on the elements. In the case of the earth, the meditator is advised to gather together a lump of "pure dawn-colored clay," somewhere in size between a saucer and a bushel in diameter, and shape it into the form of an even disk.[1] One is then advised to gaze at the earth mound for an extended period of time, reciting various epithets for the earth: *pathavī /pṛthivī*, an ancient term used to refer to the Earth Goddess; *mahī*, the Great One; *medinī*, the Friendly One; *bhūmi*, the Ground; *vasudhā*, the Provider of Wealth; and *vasudharā*, the Bearer of Wealth.[2] In this process of concentration, the body is held stable in a meditative pose; the sense of sight is restricted to the mound of earth; the mind is held in check by repeating various terms associated with earth. After repeated practice with eyes open, one then masters the ability to visualize the earth *kasina* with eyes closed. At this point, one enters into the "production phase" and is able to recreate this meditation regardless of place or circumstance.

This process is then followed in succession by the water, fire, air, and space *kasinas*. For cultivating concentration on water, one is to place clear water in a bowl and gaze upon it, using the words water (*āpo*), rain (*ambu*), liquid (*udaka*), dew (*vāri*), and fluid (*salila*).[3] For concentration on fire, one creates a wood fire and then sets the mind on a variety of its appelations: *agni*, *tejo*, the Bright One (*pāvaka*), etc.[4] For the air *kasina*, one is advised as follows:

> One who is learning the air kasina apprehends the signs in air . . .
> notices the tops of growing sugarcane moving to and fro, . . . or
> the tops of bamboos, or the tops of trees, or the ends of the hair,
> moving to and fro; or notices the touch of it on the body.[5]

The words associated with air that help focus one's attention include wind (*vāta*), breeze (*māluta*), and blowing (*anila*). One then meditates progressively on the colors blue, yellow, red, and white, using flowers or cloth as a support, and then turns to light "thrown on a wall or a floor by sunlight or moonlight."[6] The final *kasina* involves meditation on space, noticed when one sees "a hole in a wall, or in a keyhole, or in a window opening."

When each of these is cultivated, benefits are said to accrue: "The hindrances eventually become suppressed, the defilements subside, and the mind becomes concentrated."[7] This power wears away conventional constructs of the mind that see things in their particularity rather than in their shared essence. In my own direct experience of cultivating the earth concentration, all sorts of material objects, from dishes to automobiles to electronic equipment, appear to be none other than variations on the same primal matter: different configurations and usages of Mother Earth. This perception helps to hold back the tendency to send the five senses outward (*prapañca*); one does not see worldly objects as inherently desirable, but sees them in terms of their root origins. Pushing the meditation a bit further, one sees that their arising cannot take place without the power of intentional awareness. Once seeing that an object depends radically upon the subject's own structures of mind, the sending forth of that mind to the object becomes optional. In this way, hindrances of karma can be reversed. The mind and its experience of and relation to things becomes purified.

One important point to bear in mind when discussing meditation on the successive components of nature is that this approach finds its ground in concrete realities. This is not a Romanticism where the landscape becomes idealized and abstract. Beautiful vistas are not required; it can be practiced within the confines of one's own home. However, the usefulness of directing one's attention to natural beauty cannot be denied, even in the case of persons whose intentions are not necessarily meditational or religious. Throughout the Sierra Pelona Mountains of northern Los Angeles County are located several youth detention camps, far removed from the city below, where juvenile offenders work off their sentencing by clearing brush and maintaining trails. The young men and women assigned to this duty as punishment for their wrongdoings have a far better rate of non-return than those who are conventionally incarcerated. The power of the land speaks and helps heal them.

The women of the Chipko movement in India have expressed their intimacy with the earth in language that is simple yet compelling. When asked to explain the passion of the women who risk their lives to spare India's remaining stands of forest, Vandana Shiva replies:

> All the nature deities are always female, by and large, because all of them are considered *prakṛti*, the female principle in Hindu cosmology. Also, all of them are nurturing mothers—the trees feed you, the streams feed you, the land feeds you, and everything that nurtures you is a mother.[8]

This force of *prakṛti* or female creativity goes to the very core of India's spirituality. The power of the feminine has consistently spoken through India's prehistorical figurines of the feminine, through the goddess-ification of her rivers, the dawn, and the power of speech in the *Rig Veda*, to her gallant tales of strong women such as Draupadī and Sītā and such goddesses as Sarasvati, Lakṣmi, and Kālī.

By the time of the seventh or eighth century of the common era, an intellectual system arose stemming from this early view of the universe and body that came to profoundly influence all three of India's indigenous traditions: Hinduism, Jainism, and Buddhism. This system, known as tantra, divided reality into two primary components: the female force or *śakti* and the male force or *śaktā*. These two interact in such a way that the universe unfolds continually through a process of vibration (*spanda*), the power of which is captured in root syllables or chants (*mantra*) that can both awaken the goddess within the various energy centers (*cakras*) of the body and the elements associated with these centers out of which our own bodies are formed.

Environmental activism is closely linked to bodily awareness. India's spiritual traditions, including devotion to the feminine and sophisticated techniques of meditation on the essential elements of nature, provide rich resources for persons reconnecting with our bodies and with the world. By learning from some of these insights and practices, we can begin to live with the sensitivity, sensibility, and frugality that the newly emerging ecozoic era demands.

—19—
Spirit in Action

In Service of the Earth and Humanity

Angeles Arrien

Move from within. Don't move the way that fear wants you to.
Begin a foolish project. Noah did.

—Rumi

Native peoples of the world hold the belief that we are all original medicine, however else duplicated in the world. In these times of change, we are being called on to employ our creative fire and to bring forward those "foolish projects," our life's dreams. It is very important for us to take our place to shift the shape of our reality or our experience, and to build new worlds both internally and externally.

An example of this occurred at Clinton's Economic Summit. A Navajo elder, who was the spokesperson for all the Native peoples on reservations, came forward to speak. He came to talk about the plight of those living on the reservations in this country. He acknowledged first, with gratitude, all of the people who were there. He then proceeded to describe what was working on the reservations. After that he acknowledged what was not working. He concluded by saying, "I regret that I have only three creative solutions to offer."

What a healing experience it would be for all of us to use this model of communication in our families, in our professional lives, and in our spiritual explorations. If we could start with acknowledging those present, and then say what is working, the next stop would be to address what isn't working. Finally the last step would be to generate at least *three* creative solutions to the problem, not just one solution to which we have become attached.

It is very easy to complain—and it is important to acknowledge what is not working. However, it is time for us to become creative catalysts and healing agents. As catalysts, we need to bring forward creative solutions and to master what Native peoples of the world call *shape-shifting*. We must stop "describing" what is happening—and move instead to "prescribing"; to move out of description and into prescription.

Creating Small Miracles: Milagros Pequeños

Shape-shifting is an opportunity presented to us every day. Recently while I was waiting for the shuttle to take me to the airport, I was sitting on the bench with another woman, who was reading a newspaper. A fourteen-year-old boy, with his baseball cap on backwards, was buzzing us on his skateboard. He buzzed us once, and then buzzed us again. The third time, he came inadvertently too close, and knocked the newspaper out of the woman's hand.

Startled, she said, "Oh, why don't you grow up?" The boy skated off, down to the corner to talk to his buddy, and the two of them looked back at us. The woman then picked up her newspaper and walked to the middle of the block.

She called to him, "Could I talk to you for a moment?" Reluctantly, the boy came back, very slowly, on his skateboard. He stopped and took hold of his hat and turned it around in front, with the bill up and said, "Yeah?" She said, "What I meant to say . . . what I meant to say was that I was afraid that you might hurt me, and I apologize for saying what I did."

He looked at her, his face lit up, and he said, "How cool!" That moment is indelibly etched in my mind.

What touched me most was that within the space of five minutes, a human being decided to shape-shift the experience and create what in Latin America is called a *milagro pequeño*, a small miracle, a holy moment. This was an experience created between a boy and a woman, generations apart: a small miracle. We have the opportunity, as creative catalysts and healing agents, to become shape-shifters to create many holy moments, many *milagro pequeños*.

While addressing the Elmwood Institute, Vice-President Al Gore reminded us that these holy moments heed the essential call of Spirit in the twenty-first century when he said: "The more deeply I search for the root of the global environmental crisis, the more I am convinced that it is an outer manifestation of an inner crisis that is, for lack of a better word, spiritual."

Nature and Spirit in Ancient and Modern Times

Indigenous peoples of each continent recognize the deep connection between Nature and Spirit. These land-based cultures hold the technology and understanding of how to survive in nature. Prevalent in indigenous land-based wisdoms and in modern technology are these pertinent questions: How can we bridge and integrate the old and new without sacrificing one for the other? How can we use modern fiber optic and computer technology to support an understanding and appreciation of nature, and help solve our current environmental issues? Our task is how to employ an unobtrusive and progressive technology and still maintain a world environment that is left "natural" and unharmed.

Two post-modern spiritual world views of "green ecology" and the "New Age" are contemporary responses to bridging nature and spirit. Today it is imperative that we pay attention to ecological issues. Our planet, the house we live in, is in danger of becoming unlivable, primarily due to the neglect of our own industrialized society. It is clear that we need to take action for change before it is too late.

As we move into the twenty-first century, it is the work of all human beings to attend to the health of both our "inner" and "outer" houses: the inner house or ourselves, the limitless world within; and the outer house of the world in which we live our daily lives. Many people in contemporary society feel little or no connection between those two worlds. This is a state that the indigenous, land-based peoples of the earth, whose cultures reach back thousands of years, would find not only sad but incomprehensible.

In *Voices of the First Day*, Robert Lawlor quotes an Aboriginal Tribal Elder saying, "They say we have been here for 60,000 years, but it is much longer. We have been here since the time before time began. We have come directly out of the Dreamtime of the Creative Ancestors. We have lived and kept the earth as it was on the First Day." It is to Native peoples that we can now look for guidance on how to care for the earth as if it were its first day. Their universal, ancient wisdoms can help restore balance within our own nature, and assist in rebalancing the needs of the natural environment.

An Integrated Vision: The Braided Way

The United Nations declared 1993 as the Year for Honoring Indigenous Peoples of the World. It is crucial that we support and preserve the wisdom of Nature found among indigenous cultures and contemporary farm and husbandry societies. These land-based wisdoms can support our continued stewardship and survival of Mother Nature.

During cycles of change and challenge, we can become more effective by holding the vision and possibility of a "braided way." A braid can be a symbol for unifying polarities of any kind, creating a third option. For example, one portion of the braid could represent the new, one portion the old. The third portion of the braid is where the old and the new are equally honored, creating a neutral ground where similarities and differences can co-exist and support one another.

We live in an age that is calling for a "new world order." Our current world order actually consists of four worlds: (1) the highly industrialized First World countries, such as the United States and the nations of Western Europe; (2) the Second World socialist block nations; (3) the developing countries of the Third World, such as Brazil or Thailand; and (4) the Fourth World, which George Manuel of the World Council of Indigenous Peoples describes in *The Gaia Atlas of First Peoples* (Burger) as the "name given to indigenous peoples descended from a country's aboriginal population and who today are completely or partially deprived of the right to their own territory and riches. The peoples of the Fourth World have only limited influence or none at all in the national state to which they belong."

The differences between these worlds can be stated very simply: The First, Second and Third Worlds believe that "the land belongs to the people"; the Fourth World believes that "the people belong to the land." An integrative model can be created once all four worlds create a bridge that is truly healing. Perhaps this bridge can be the interface or braided way where these four worlds meet in joining together to heal and restore Mother Earth.

For the people of the first three worlds, understanding and accepting the belief of the Fourth World is the first and most crucial step in creating a truly new and interconnected world. This may seem impossible, but it is not. The interface between the worlds is not rigid and impenetrable. Noted psychologist William Bridges explains the meaning of interface in his book *Surviving Corporate Transition*:

> . . . where the surface of one thing meets the surface of another. It is less like a dividing line and more like a permeable membrane, and the action at the interface is the interplay, the communication, the mutual influence that goes on between societies . . . that are side by side. The interface is where the vital relationships are established that are necessary for survival in a world of increasing interdependency. [1]

No matter what world we live in now, we are all people of the earth, connected to one another by our mutual humanity. When we listen to land-based peoples, we are listening to our oldest selves. When we listen to our progressive technology, we are listening to our creative selves. Together they can facilitate Spirit in action in co-creating a new course for the world.

Let us consciously honor and respect Native peoples and Mother Nature. Let us use our progressive technology to support humanity and

Mother Earth. It is very important in these times for us to become stewards of our inner and outer lives, to become shape-shifters of our experience, to honor the braided way, and lastly, to create many *milagros pequeños*. In this way, we can honor the vision of what I refer to as the Four-Fold Way, by collectively showing up (or choosing to be present), paying attention to what has heart and meaning, telling the truth without blame or judgment, and being open to outcome, not attached to outcome. When we do this, we enhance the survival and evolution of Mother Earth by creating the foundations for a possible new world.

People, Land, and Community

Wendell Berry

I would like to speak more precisely than I have before of the connections that join people, land and community—to describe, for example, the best human use of a problematical hillside farm. In a healthy culture, those connections are complex. The industrial economy breaks them down by oversimplifying them and in the process raises obstacles that make it hard for us to see what the connections are or ought to be. These are mental obstacles, of course, and there appear to be two major ones: the assumption that knowledge (information) can be "sufficient," and the assumption that time and work are short.

These assumptions will be found implicit in a whole set of contemporary beliefs: that the future can be studied and planned for; that limited supplies can be wasted without harm; that good intentions can safeguard the use of nuclear power. A recent newspaper article says, for example, "A congressionally mandated study of the Ogallala Aquifer is finding no great cause for alarm from [*sic*] its rapidly dropping levels. The director of the . . . study . . . says that even at current rates of pumping, the aquifer can supply the Plains with water for another forty to fifty years. . . . All six states participating in the study . . . are forecasting increased farm yields based on improved technology." Another article speaks of a different technology with the same optimism: "The nation has invested hundreds of billions of dollars in atomic weapons and at the same time has developed the most sophisticated strategies to fine-tune their use to avoid a holocaust. Yet the system that is meant to activate them is the weakest link in the chain. . . . Thus, some have suggested that what may be needed are warning systems for the warning systems."

Always the assumption is that we can first set demons at large, and then, somehow, become smart enough to control them. This is not childishness. It is not even "human weakness." It is a kind of idiocy, but perhaps we will not cope with it and save ourselves until we regain the sense to call it evil.

The trouble, as in our conscious moments we all know, is that we are terrifyingly ignorant. The most learned of us are ignorant. The acquisition of knowledge always involves the revelation of ignorance—almost *is* the revelation of ignorance. Our knowledge of the world instructs us first of all that the world is greater than our knowledge of it. To those who rejoice in the abundance and intricacy of Creation, this is a source of joy, as it is to those who rejoice in freedom. ("The future comes only by surprise," we say, "—thank God!") To those would-be solvers of "the human problem," who hope for knowledge equal to (capable of controlling) the world, it is a source of unremitting defeat and bewilderment. The evidence is overwhelming that knowledge does not solve "the human problem." Indeed, the evidence overwhelmingly suggests—with Genesis—that knowledge *is* the problem. Or perhaps we should say instead that all our problems tend to gather under two questions about knowledge: Having the ability and desire to know, how and what should we learn? And, having learned, how and for what should we use what we know?

One thing we do know, that we dare not forget, is that better solutions than ours have at times been made by people with much less information than we have. We know too, from the study of agriculture, that the same information, tools, and techniques that in one farmer's hands will ruin land, in another's will save and improve it.

This is not a recommendation of ignorance. To know nothing, after all, is no more possible than to know enough. I am only proposing that knowledge, like everything else, has its place, and that we need urgently now to *put* it in its place. If we want to know and cannot help knowing, then let us learn as fully and accurately as we decently can. But let us at the same time abandon our superstitious beliefs about knowledge: that it is ever sufficient; that it can of itself solve problems; that it is intrinsically good; that it can be used objectively or disinterestedly. Let us acknowledge that the objective or disinterested researcher is always on the side that pays the best. And let us give up our forlorn pursuit of the "informed decision."

The "informed decision," I suggest, is as fantastical a creature as the "disinterested third party" and the "objective observer." Or it is if by "informed" we mean "supported by sufficient information." A great deal of our public life, and certainly the most expensive part of it, rests on the assumed possibility of decisions so informed. Examination of private life, however, affords no comfort whatsoever to that assumption. It is simply true that we do not and cannot *know* enough to make any important decision.

Of this dilemma we can take marriage as an instance, for as a condition marriage reveals the insufficiency of knowledge, and as an institution it suggests the possibility that decisions can be informed in another way that *is* sufficient, or approximately so. I take it as an axiom that one cannot know enough to get married, any more than one can predict a surprise. The only people who possess information sufficient to their vows are widows and widowers—who do not know enough to *re*marry.

What is not so well understood now as perhaps it used to be is that marriage is made in an inescapable condition of loneliness and ignorance, to which it, or something like it, is the only possible answer. Perhaps this is so hard to understand now because now the most noted solutions are mechanical solutions, which are often exactly suited to mechanical problems. But we are humans—which means that we not only *have* problems but *are* problems. Marriage is not as nicely trimmed to its purposes as a bottle-stopper; it is a not entirely possible solution to a not entirely soluble problem. And this is true of the other human connections. We can commit ourselves fully to anything—a place, a discipline, a life's work, a child, a family, a community, a faith, a friend—only in the same poverty of knowledge, the same ignorance of result, the same self-subordination, the same final forsaking of other possibilities. If we must make these so final commitments without sufficient information, then what *can* inform our decisions?

In spite of the obvious dangers of the word, we must say first that love can inform them. This, of course, though probably necessary, is not safe. What parent, faced with a child who is in love and going to get married, has not been filled with mistrust and fear—and justly so. We who were lovers before we were parents know what a fraudulent justifier love can be. We know that people stay married for different reasons than those for which they got married and that the later reasons will have to be discovered. Which, or course, is not to say that the later reasons may not confirm the earlier ones; it is to say only that the earlier ones must wait for confirmation.

But our decisions can also be informed—our loves both limited and strengthened—by those patterns of value and restraint, principle and expectation, memory, familiarity, and understanding that, inwardly, add up to *character* and, outwardly, to *culture*. Because of these patterns, and only because of them, we are not alone in the bewilderments of the human condition and human love, but have the company and the comfort of the best of our kind, living and dead. These patterns constitute a knowledge far different from the kind I have been talking about. It is a kind of knowledge that includes information, but is never the same as information. Indeed, if we study the paramount documents of our culture, we will see that this second kind of knowledge invariably implies, and often explicitly imposes, limits upon the first kind: some possibilities must not be explored; some things must not be learned. If we want to get safely home, there are certain seductive songs we must not turn aside for, some sacred things we must not meddle with:

> Great captain,
> a fair wind and the honey lights of home
> are all you seek. But anguish lies ahead;
> the god who thunders on the land prepares it . . .
>
> . . . .
>
> One narrow strait may take you through his blows:
> denial of yourself, restraint of shipmates.

This theme, of course, is dominant in biblical tradition, but the theme itself and its modern inversion can be handily understood by a comparison of this speech of Tiresias to Odysseus in Robert Fitzgerald's Homer with Tennyson's romantic Ulysses who proposes, like a genetic engineer or an atomic scientist,

> To follow knowledge like a sinking star,
> Beyond the utmost bound of human thought.

Obviously unlike Homer's Odysseus, Tennyson's Ulysses is said to come from Dante, and he does resemble Dante's Ulysses pretty exactly—the critical difference being that Dante thought this Ulysses a madman and a fool, and brings down upon his Tennysonian speech to his sailors one of the swiftest anticlimaxes in literature. The real—the human—knowledge is understood as implying and imposing limits, much as marriage does, and these limits are understood to belong necessarily to the definition of a human being.

In all this talk about marriage I have not forgot that I am supposed to be talking about agriculture. I am going to talk directly about agriculture in a minute, but I want to insist that I have been talking about it indirectly all along, for the analogy between marriage making and farm making, marriage keeping and farm keeping, is nearly exact. I have talked about marriage as a way of talking about farming because marriage, as a human artifact, has been more carefully understood than farming. The analogy between them is so close, for one thing, because they join us to time in nearly the same way. In talking about time, I will begin to talk directly about farming, but as I do so, the reader will be aware, I hope, that I am talking indirectly about marriage.

When people speak with confidence of the longevity of diminishing agricultural sources—as when they speak of their good intentions about nuclear power—they are probably not just being gullible or thoughtless;

they are likely to be speaking from belief in several tenets of industrial optimism: that life is long, but time and work are short; that every problem will be solved by a "technological breakthrough" before it enlarges to catastrophe; that *any* problem can be solved in a hurry by large applications of urgent emotion, information, and money. It is regrettable that these assumptions should risk correction by disaster when they could be cheaply and safely overturned by the study of any agriculture that has proved durable.

To the farmer, Emerson said, "The landscape is an armory of powers. . . . " As he meant it, the statement may be true, but the metaphor is ill-chosen, for the powers of a landscape are available to human use in nothing like so simple a way as are the powers of an armory. Or let us say, anyhow, that the preparations needed for the taking up of agricultural powers are more extensive and complex than those usually thought necessary for the taking up of arms. And let us add that the motives are, or ought to be, significantly different.

Arms are taken up in fear and hate, but it has not been uncharacteristic for a farmer's connection to a farm to begin in love. This has not always been so ignorant a love as it sometimes is now; but always, no matter what one's agricultural experience may have been, one's connection to a newly bought farm will begin in love that is more or less ignorant. One loves the place because present appearances recommend it, and because they suggest possibilities irresistibly imaginable. One's head, like a lover's, grows full of visions. One walks over the premises, saying, "If this were in, I'd make a permanent pasture here; here is where I'd plant an orchard; here is where I'd dig a pond." These visions are the usual stuff of unfulfilled love and induce wakefulness at night.

When one buys the farm and moves there to live, something different begins. Thoughts begin to be translated into acts. Truth begins to intrude with its matter-of-fact. One's work may be defined in part by one's visions, but it is defined in part too by problems, which the work leads to and reveals. And daily life, work, and problems gradually alter the visions. It invariably turns out, I think, that one's first vision of one's place was to some extent an imposition on it. But if one's sight is clear and if one stays on and works well, one's love gradually responds to the place as it really is, and one's visions gradually image possibilities that are really in it. Vision, possibility, work, and life—*all* have changed by mutual correction. Correct discipline, given enough time, gradually removes one's self from one's line of sight. One works to better purpose then and makes few mistakes, because at last one sees where one is. Two human possibilities of the highest order thus come within reach: what one wants can become the same as what one has, and one's knowledge can cause respect for what one knows.

"Correct discipline" and "enough time" are inseparable notions. Correct discipline cannot be hurried, for it is both the knowledge of what ought to be done, and the willingness to do it—*all* of it, properly. The good worker will not suppose that good work can be made properly answerable to haste, urgency, or even emergency. But the good worker knows too that after it is done work requires yet more time to prove its worth. One must stay to experience and study and understand the consequences—must understand them by living with them, and then correct them, if necessary, by longer living and more work. It won't do to correct mistakes made in one place by moving to another place, as has been the common fashion in America, or by adding on another place, as is the fashion in any sort of "growth economy." Seen this way, questions about farming become inseparable from questions about propriety of scale. A farm can be too big for a farmer to husband properly or pay proper attention to. Distraction is inimical to correct discipline, and enough time is beyond the reach of anyone who has too much to do. But we must go farther and see that propriety of scale is invariably associated with propriety of another kind: an understanding and acceptance of the human place in the order of Creation—a proper humility. There are some things the arrogant mind does not see; it is blinded by its vision of what it desires. It does not see what is already there; it never sees the forest that precedes the farm or the farm that preceded the shopping center; it will never understand that America was "discovered" by the Indians. It is the properly humbled mind in its proper place that sees truly, because—to give only one reason—it sees details.

And the good farmer understands that further limits are imposed upon haste by nature which, except for an occasional storm or earthquake, is in no hurry either. In the processes of most concern to agriculture—the building and preserving of fertility—nature is never in a hurry. During the last seventeen years, for example, I have been working at the restoration of a once exhausted hillside. Its scars are now healed over, though still visible, and this year it has provided abundant pasture, more than in any year since we have owned it. But to make it as good as it is now has taken seventeen years. If I had been a millionaire or if my family had been starving, it would still have taken seventeen years. It can be better than it now is, but that will take longer. For it to live fully in its own possibility, as it did before bad use ran it down, may take hundreds of years.

But to think of the human use of a piece of land as continuing through hundreds of years, we must greatly complicate our understanding of agriculture. Let us start a job of farming on a given place—say an initially fertile hillside in the Kentucky River Valley—and construe it through time:

1. To begin using this hillside for agricultural production—pasture or crop—is a matter of a year's work. This is work in the present tense, adequately comprehended by conscious intention and by the first sort of knowledge I talked about—information available to the farmer's memory and built into his methods, tools, and crop and livestock species. Understood in its present tense, the work does not reveal its value except insofar as the superficial marks of craftsmanship may be seen and judged. But excellent workmanship, as with a breaking plow, may prove as damaging as bad workmanship. The work has not revealed its connections to the place or to the worker. These connections are revealed in time.

2. To live on the hillside and use if for a lifetime gives the annual job of work a past and a future. To live on the hillside and use it without diminishing its fertility or wasting it by erosion still requires conscious intention and information, but now we must say *good* intention and *good* (that is, correct) information, resulting in *good* work. And to these we must now add *character*: the sort of knowledge that might properly be called familiarity, and the affections, habits, values, and virtues (conscious and unconscious) that would preserve good care and good work through hard times.

3. For human life to continue on the hillside through successive generations requires good use, good work, all along. For in any agricultural place that will waste or erode—and all will—bad work does not permit "muddling through"; sooner or later it ends human life. Human continuity is virtually synonymous with good farming, and good farming obviously must outlast the life of any good farmer. For it to do this, in addition to the preceding requirements, we must have *community*. Without community, the good work of a single farmer or a single family will not mean much or last long. For good farming to last, it must occur in a good farming community—that is, a neighborhood of people who know each other, who understand their mutual dependencies, and who place a proper value on good farming. In its cultural aspect, the community is an order of memories preserved consciously in instructions, songs, and stories, and both consciously and unconsciously in *ways*. A healthy culture holds preserving knowledge *in place* for a *long* time. That is, the essential wisdom accumulates in the community much as fertility builds in the soil. In both, death becomes potentiality.

People are joined to the land by work. Land, work, people, and community are all comprehended in the idea of culture. These connections cannot be understood or described by information—so many resources to be transformed by so many workers into so many products for so many consumers—because they are not quantitative. We can understand them only after we acknowledge that they should be harmonious—that a culture must be either shapely and saving or shapeless and destructive. To presume to describe land, work, people, and community by information, by quantities, seems invariably to throw them into competition with one another. Work is then understood to exploit the land, the people to exploit their work, the community to exploit its people. And then instead of land, work, people, and community, we have the industrial categories of resources, labor, management, consumers, and government. We have exchanged harmony for an interminable fuss, and the work of culture for the timed and harried labor of an industrial economy.

But let me bring these notions to the trial of a more particular example.

Wes Jackson and Marty Bender of the Land Institute have recently worked out a comparison between the energy economy of a farm using draft horses for most of its field work and that of an identical farm using tractors. This is a project a generation overdue, of the greatest interest and importance—in short, necessary. And the results will be shocking to those who assume a direct proportion between fossil fuel combustion and human happiness.

These results, however, have not fully explained one fact that Jackson and Bender had before them at the start of their analysis and that was still running ahead of them at the end: that in the last twenty-five or thirty years, the Old Order Amish, who use horses for farmwork, doubled their population and stayed in farming, whereas in the same period millions of mechanized farmers were driven out. The reason that this is not adequately explained by analysis of the two energy economies, I believe, is that the problem is by its nature beyond the reach of analysis of any kind. The real or whole reason must be impossibly complicated, having to do with nature, culture, religion, family and community life, as well as with agricultural methodology and economics. What I think we are up against is an unresolvable difference between thought and action, thought and life.

What works *poorly* in agriculture—monoculture, for instance, or annual accounting—can be pretty fully explained, because what works poorly is invariably some oversimplifying *thought* that subjugates nature, people, and culture. What works well ultimately defies explanation because it involved an order which in both magnitude and complexity is ultimately incomprehensible.

Here, then, is a prime example of the futility of a dependence on information. We cannot contain what contains us or comprehend what comprehends us. Yeats said that "Man can embody truth but he cannot know it." The part, that is, cannot comprehend the whole, though it can stand for it (and by it). Synecdoche is possible, and its possibility implies the possibility of harmony between part and whole. If we cannot work on the basis of sufficient information, then we have to work on the basis of an understanding of harmony. That, I take it, is what Sir Albert Howard and Wes Jackson mean when they tell us that we must study and emulate on our farms the natural integrities that precede and support agriculture.

The study of Amish agriculture, like the study of *any* durable agriculture, suggests that we live in sequences of patterns that are formally analogous. These sequences are probably hierarchical, at least in the sense that some patterns are more comprehensive than others; they tend to arrange themselves like interesting bowls—though any attempt to represent their order visually will oversimplify it.

And so we must suspect that Amish horse-powered farms work well, not because—or not *just* because—horses are energy-efficient, but because they are living creatures, and therefore fit harmoniously into a pattern of relationships that are necessarily geological, and that rhyme analogically from ecosystem to crop, from field to farmer. In other words, ecosystem, farm, field, crop, horse, farmer, family, and community are in certain critical ways *like* each other. They are, for instance, all related to health and fertility or reproductivity in about the same way. The health and fertility of each involves and is involved in the health and fertility of all.

It goes without saying that tools can be introduced into this agriculture and ecological order without jeopardizing it—but only up to a certain kind, scale, and power. To introduce a tractor into it, as the historical record now seems virtually to prove, is to begin its destruction. The tractor has been so destructive, I think, because it is *unlike* anything else in the agriculture order, and so it breaks the essential harmony. And with the tractor comes dependence on any energy supply that lies not only off the farm but outside agriculture and outside biological cycles and integrities. With the tractor, both farm and farmer become "resources" of the industrial economy, which always exploits it resources.

We would be wrong, of course, to say that anyone who farms with a tractor is a bad farmer. That is not true. What we must say, however, is that once a tractor is introduced into the pattern of a farm, certain necessary restraints and practices, once implicit in technology, must now reside in the character and consciousness of the farmer—at the same time that the economic pressures to cast off restraint and good practice has been greatly increased.

In a society addicted to facts and figures, anyone trying to speak for agricultural *harmony* is inviting trouble. The first trouble is in trying to say what harmony is. It cannot be reduced to facts and figures—though the lack of it can. It is not very visibly a function. Perhaps we can only say what it may be like. It may, for instance, be like sympathetic vibration: "The A string of a violin . . . is designed to vibrate most readily at about 440 vibrations per second: the note A. If that same note is played loudly not on the violin but near it, the violin A string may hum in sympathy." This may have a practical exemplification in the craft of the mud daubers which, as they trowel mud into their nest walls, hum to it, or at it, communicating a vibration that makes it easier to work, thus mastering their material by a kind of song. Perhaps the hum of the mud dauber only activates that anciently perceived likeness between all creatures and the earth of which they are made. For as common wisdom holds, like *speaks* to like. And harmony always involves such specificities of form as in the mud dauber's song and its nest, whereas information accumulates indiscriminately, like noise.

Of course, in the order of creatures, humanity is a special case. Humans, unlike mud daubers, are not naturally involved in harmony. For humans, harmony is always a human product, an artifact, and if they do not know how to make it and choose to make it, then they do not have it. And so I suggest that, for humans, the harmony I am talking about may bear an inescapable likeness to what we know as moral law—or that, for humans, moral law is a significant part of the notation of ecological and agriculture harmony. A great many people seem to have voted for information as a safe substitute for virtue, but this ignores—among much else—the need to prepare humans to live short lives in the face of long work and long time.

Perhaps it is only when we focus our minds on our machines that time seems short. Time is always running out for machines. They shorten our work, in a sense popularly approved, by simplifying it and speeding it up, but our work perishes quickly in them too as they wear out and are discarded. For the living Creation, on the other hand, time is always coming. It is running out for the farm built on the industrial pattern; the industrial farm burns fertility as it burns fuel. For the farm built into the pattern of living things, as an analogue of forest or prairie, time is a bringer of gifts. These gifts may be welcomed and cared for. To some extent they may be expected. Only within strict limits are they the result of human intention and knowledge. They cannot in the usual sense be made. Only in the short term industrial accounting can they be thought simply earnable. Over the real length of human time, to be earned they must be deserved.

From this rather wandering excursion I arrive at two conclusions.

The first is that the modern stereotype of an intelligent person is probably wrong. The prototypical modern intelligence seems to be that of the

Quiz Kid—a human shape barely discernible in fluff of facts. It is understood that everything must be justified by facts, and facts are offered in justification of *everything*. If it is a fact that soil erosion is now a critical
problem in American agriculture, then more facts will indicate that it is not
as bad as it *could* be and that Iowa will continue to have topsoil for as long
as seventy more years. If facts show that some people are undernourished in
America, further facts reveal that we should all be glad we do not live in
India. This, of course, is machine thought.

To think better, to think like the best humans, we are probably going to
have to learn again to judge a person's intelligence, not by the ability to recite
facts, but by the good order or harmoniousness of his or her surroundings.
We must suspect that any statistical justification of ugliness and violence is a
revelation of stupidity. As an earlier student of agriculture put it: "The intelligent man, however unlearned, may be known by his surroundings, and by
the care of his horse, if he is fortunate enough to own one."

My second conclusion is that any public program to preserve land or
produce food is hopeless if it does not tend to right the balance between
numbers of people and acres of land, and to encourage long-term, stable connections between families and small farms. It could be argued that our
nation has never made an effort in this direction that was knowledgeable
enough or serious enough. It is certain that no such effort, here, has ever
succeeded. The typical American farm is probably sold and remade—often
as part of a larger farm—at least every generation. Farms that have been
passed to the second generation of the same family are unusual. Farms that
have passed to the third generation are rare.

But our crying need is for an agriculture in which the typical farm
would be farmed by the third generation of the same family. It would be
wrong to try to say exactly what kind of agriculture that would be, but it
may be allowable to suggest that certain good possibilities would be
enhanced.

The most important of those possibilities would be the lengthening of
memory. Previous mistakes, failures, and successes would be remembered.
The land would not have to pay the cost of a trial-and-error education for
every new owner. A half century or more of the farm's history would be living memory, and its present state of health could be measured against its
own past—something exceedingly difficult *outside* of living memory.

A second possibility is that the land would not be overworked to pay
for itself at full value with every new owner.

A third possibility would be that, having some confidence in family
continuity in place, present owners would have future owners not only in

supposition but *in sight* and so would take good care of the land, not for the sake of something so abstract as "the future" or "posterity," but out of particular love for living children and grandchildren.

A fourth possibility is that having the past so immediately in memory, and the future so tangibly in prospect, the human establishment on the land would grow more permanent by the practice of better carpentry and masonry. People who remembered long and well would see the folly of rebuilding their barns every generation or two, and of building new fences every twenty years.

A fifth possibility would be the development of the concept of *enough*. Only long memory can answer, for a given farm or locality, How much land is enough? How much work is enough? How much livestock and crop production is enough? How much power is enough?

A sixth possibility is that of local culture. Who could say what that would be? As members of a society based on the exploitation of its own temporariness, we probably should not venture a guess. But we can perhaps speak with a little competence of how it would begin. It would not be imported from critically approved cultures elsewhere. It would not come from watching certified classics on television. It would begin in work and love. People at work in communities three generations old would know that their bodies renewed, time and again, the movements of other bodies, living and dead; known and loved, remembered and loved, in the same shops, houses, and fields. That, of course, is a description of a kind of community dance. And such a dance is perhaps the best way we have to describe harmony.

—21—
On Savage Art

James Cowan

True culture operates by exaltation and force.

—Antonin Artaud

Deep in the central desert region of Australia a number of nomad tribes are beginning the long road back to reclaiming their culture by way of painting works of art on canvas. It is a recent phenomenon, going back little more than twenty-five years. Tribal elders who once roamed the desert in search of food and spiritual nourishment are now sitting on the ground, a stretched piece of canvas before them, painting in rich ochre colors the contours of their spirit-country, their visionary realm. They have found a unique method by which they can once more forge links with old memories, with long-abandoned rituals, and with the songs and dances of their ancestral past. It is a moment of supreme importance for them since they are now able to recreate the lineaments of their savage self once more.

The idea that we might be inspired by a sense of wildness has always fascinated me. For many of us savagery suggests cruelty and barbarism, the very antipathy of civilized virtues. Refined cultures such as ours thrive on values borne out of intellectual processes rather than natural ones. We have buttressed ourselves against the very forces that made us, hoping to contain them. We do not wish to be at the mercy of those great, free forces of nature which owe allegiance to no one but themselves. Nature's absolute power causes us to tremble even when we choose to harness it. An atomic mushroom cloud, or a torrent surging forth from the base of a dam, strike us as a power bridled yet somehow resistant to our control. For a brief moment we might believe we have shackled its energy, but in our hearts we know this is not so. Somehow nature's primordial savagery, its wildness, remains bound like a prisoner in the dungeon of our psyche, awaiting its moment to break free.

I say these things in the light of my yearly encounters with Aborigines. At first I always felt slightly uncomfortable in their presence, as if a certain mode of contact, though implicit, had not been achieved. It was as if we were two distinct species, a dog and cat perhaps, who were meeting for the first time. We gazed at each other, observed our mannerisms, but still we did not know how to react. Neither of us had learned how to bridge the gap between our respective notions of personhood as we each harbored a belief in the supremacy of what we stood for. Here we were, caught in the net of our inherited beliefs, unable to untangle ourselves. We could not recognize the point of our common humanity. We were still seeing one another as the "other," the alien, the interloper, the primitive.

This lack of comprehension may have been comical if it were not tragic. Later I realized how this inability to recognize the essential humanness in one another signaled the birth of racial hatred and inferiority. By not confronting with equanimity the principal of diversity that had made us what we are, we found ourselves in no position to grapple with our inherent sameness. I should be more explicit: I think my Aboriginal friends knew how to handle our diversity more than I did. For they were privy to a way of thinking which allowed them to embrace the mystery of metamorphosis whereby a person could become something other than himself. Their long experience in the desert had taught them that a man could embody a spirit even as he embodied a rock or a hill. This they did through their identification with a totem.

The origin of totemic identity lies in an understanding of Creation, and how humankind was formed. In the far distant past, the *alchira*, two self-existent beings, known as Numbakulla, noticed on the eastern horizon a number of *inapertwa* or rudimentary humans who possessed neither arms nor legs, nor did they eat. The *inapertwa* represented a transitionary stage in the transformation of animals and plants into humans. Such a condition of shapelessness, when a person is regarded as little more than a red pebble among many others, is regarded by the Aranda tribe of central Australia as *kuruna*, a term that suggests man's pre-existent form, his archetype. When the Numbakulla completed their transformation, presenting them with their duality as sexual beings, they continued to retain qualities of their earlier existence. This meant that each person was intimately related to his pre-existent form, whether it were flora, fauna, or some other natural phenomenon such as fire or wind. In the here-and-now a person cannot discard his totem, because to do so is to deny what he was at the time of the Dreaming.

Totemic identity suggests that a person is both "him/herself" in one sense, but is also "another" in the sense that he (or she) participates in an earlier, transitionary form of existence. A totem becomes a mirror in which a person can see reflected back his or her ideal form prior to its manifestation as oneself. Imbedded in the idea of totemic identity is the primordial encounter between the unmanifest principle and the realm of manifestation in the guise of duality. A person only becomes "oneself" at the moment when he or she detaches from his or her ideal state (as a transitional type in the Dreaming) and puts on the garb of conditional existence. The totem acts as an Ariadne's thread which allows a person to find the way back to pre-conscious existence in the Dreaming. One finds oneself forever linked to one's own origins, both as a spiritual being and as one of nature's manifestations in the form of their totem.

Learning to understand the concept of totemic identity enabled me to begin a dialogue with Aborigines. I soon discovered that all communication between them is presaged by paying heed to the spirits of the land. They call these up as easily as we might note a change in the weather. It is not for nothing that they begin any conversation with reference to their Dreaming terrain, that imaginal land in which they were conceived. From then on one knows that such people cannot be identified except as an extension to their country, to the spirits which inhabit their Dreaming place. I had learned my first lesson about the true nature of savagery: that it derives its seminal power, its *kurunba* or telluric energy, from the land itself, and from one's totem. What is savage in a person is the invisible, throbbing, changeless presence of the Dreaming heroes who inhabit that pleromic splendor of what can only be described as a visionary landscape.

The condition of savagery has nothing to do with barbarism, or indeed any diminution of sensibility. It is, rather, a condition inspired by a sense of belonging to the earth, to being a member of an audience in attendance in nature's amphitheater. Aborigines are past masters at observing the rituals nature enacts, and in turn re-creating these for themselves as an act of mimesis. Every act for them becomes a re-enactment of the primordial moment of world-beginning when the Dreaming heroes invoked matter out of the immateriality of substance, form out of chaos. Their dances, songs, cleansing rituals, rites of passage, and ceremonies of mourning—all of these have been derived from the Promethian gesture of the Dreaming heroes. Nothing that the Aborigines do or say is not as a result of this tranformative moment in the creation of the world.

The events of the Dreaming, this recitation of world-creation, did not happen in time, nor can it be explained in terms of a physical apotheosis. Rather, when one reads between the lines, one begins to recognize that the Dreaming is an expectation, a hope, a willingness to embrace the prospect of an imaginal event being inspired by a meta-historical act. The Dreaming happened, and is happening *in their minds*. It is an ever re-occurring event that raises the believer into a realm of absolute assurance about his or her role in the fertile moment, in the Dreaming itself. Every Aborigine is thus a participant, an observer, and an orchestrator of such events. They become, in a sense, demi-urgic in their powers to ever create, ever renew, and to ever interpret the Dreaming reality for themselves.

So my understanding of the meaning of savagery deepened. Years later, when I had long begun to feel at ease in the company of desert nomads, I too began to experience the rebirth of my own savage self. It emerged from the earth when we were sitting together; it swooped on me from above when I happened to draw near to an ancient tree whose origin was in the Dreaming; it gazed at me from a rock pool, predatory, patient, hungry for my soul. Whenever I was in the company of Aborigines I always felt this "other self" pressing back the curtains of civility which governed my normal daily life. And then I felt the warmth of this savagery, this wildness, stream-ing in upon me, causing that deeper husk of experience lying dormant within to begin to open like a seed pod. It was as if I were beginning to *grow* at last! It was as if something very real and very ancient had chosen to germi-nate within me for the first time. I knew then I had begun to claim a most moving inheritance—that of my essential savage self.

In this realization, this encounter with the kernel of what was my ori-gins, I began to understand the special genius which had inspired the birth of modern Aboriginal art. As steeped as these works are in body painting and the art of sand storytelling, I knew that those that were being painted by desert tribespeople derived their strength from another source entirely. These artists were once more drawing upon a reservoir of numinosity which they had suppressed in the wake of colonization, dispossession, and social disinte-gration at the hands of white men. For almost too long they have been fear-ful of revealing such a savage power to a world they felt might wish to destroy it. They knew that to unveil the mysterious luminosity of the Dreaming to an indifferent and largely incredulous world was to hasten its extinction. Now, however, something has changed.

History will say that in the early part of the 1970s a young schoolteacher name Geoff Bardon had introduced the Aborigines of the Papunya commu-nity, west of Alice Springs, to modern art materials, and that he had encour-aged them to paint forms which derived from their ritual life. He readily acknowledged that he knew little of their spiritual reality, and so was not prepared for the sudden appearance of strange symbolic forms on the canvas. Nor was he prepared for the fact that these forms contained esoteric informa-tion which had never been re-created in a public way before. Inadvertently, Bardon had opened the way for a renewal of the great icons from the Dreaming. The Aboriginal artists suddenly found themselves painting images that in the past would have remained secret, except to the initiated elders of the tribe. For some reason they had decided to unveil what was most intrinsic to themselves: the Dreaming motifs which previously had only been painted on the ground during sacred ceremonies, or etched on the *churinga* stones embodying the spiritual essence of the Dreaming heroes, or as ritual body painting which transformed a participant into an incarnation of the Dreaming hero.

Bardon had stimulated the birth of a modern form of savage art. He had asked the old men of the community whom he deeply respected to begin the process of cultural renewal. He saw it as the only way for the Aborigines to survive. If they could tap into the primordial images of the Dreaming, if they could open the hidden text of sacred law in their possession and draw forth from it its age-old, mysterious message, then his efforts would be worthwhile. The children of the men whom he was ostensibly teaching white man's education to at the local school might at last have access to a primer of far more enduring cultural significance, one bestowed upon them by their rapidly dwindling elders. For the old men were painting mystical structure and were realizing a deep, unrelinquishing need within themselves to portray the essence of their spirituality, so that the world might begin to understand who they were.

Out of this unusual collaboration a renaissance both of Aboriginality and art became possible. Old men such as Walter Tjampitjinpa, Johnny Warrangkula Tjupurrula, and Uta Uta Tjangala struggled with the new medium, convinced that if they mastered it they would open up a new form of expression for their people. They were right. Within a few years many paintings had passed through their hands en route to the major art collections of the world. For the art world had become hungry; it wanted access to the haunting imagery of the Dreaming before it was lost. Their vision, the vision of these Pintubi tribesmen had become a psychological weapon in the battle against the West's spiritual malaise, its paralysis of belief in the wake of scientism. People saw it as the last expression of a religious art which had died late in the fourteenth century in Europe. The Pintubi, it seemed, had revived a dead artform: that of rendering their spiritual belief as a singular act of adoration.

Aboriginal art had come to stay. People were crying out to be reinvigorated by an act of artistic savagery. They wanted their physical sensibilities to be metaphorically mutilated by these ancient artifacts masquerading in the guise of acrylic paint on canvas. More than anything they wanted to be moved, to find themselves spiritually displaced by a force more powerful than themselves. In a sense they wanted to be *alienated* so that they could once again begin to feel what it was like to be a wild person inhabiting a natural landscape. The Aboriginal artists of the central desert had opened

the way for them, for they had bestowed upon them, through their can-
vasses, the gift of their savage sensibility. Wild nature, raw belief, and the
tranformative power of the Dreaming hero had been made accessible for the
first time. It was here, in these canvases, that a timeless metaphysical field
had been prepared in readiness for a new growth in spiritual awareness.

I soon discovered that the inherent savagery in Aboriginal art could not
be detached from the panoply of ritual and ceremony that surrounded the
birth of its forms. As if incised into each painting, the signs and symbols
were born out of a strict coda of law, ritual, and song. What a man or
woman learned in their youth, in their transition from adolescence to adult-
hood, in their passage towards old age—these became a part of each painting that they attempted. Along the way, the artists had learned secret
chants, they had discovered esoteric information about the Dreaming, they
had been initiated into deeper levels of cultural awareness. All of this was
rendered on canvas as invisible ingredients synonymous with the paint itself.
The artists were *painting their knowledge* of the Dreaming each time they took up
a brush and depicted their Dreaming place.

I further discovered that savagery in art is also determined by the primacy of sapiential possession. An artist must know in great detail all the
sacred ramifications of the country he is depicting. Such information belongs
to him by way of being born *into* a region. Since conception and birth are
determined by the country itself, it follows that a person is a product of his
country as much as he is derived from his mother's womb. Such biological
events become less significant than the spiritual act of genesis implied by the
potency of the land. In birth a person becomes a living embodiment of
earth-knowledge, of the sapiential nature (i.e., wisdom) of his Dreaming
place. He inherits from his ancestors, and by implication the Dreaming
heroes themselves, a core of knowledge which stems from the earth. He
embodies a wildness, a savagery, even as he acknowledges his birthplace as
being a crucible of organic ferment.

The savage artist is one who acknowledges his origins in this way. He is
not afraid to spill blood, or to use it as a part of his body painting. He is
not afraid to suffer pain in the interest of spiritual transformation. He is not
afraid to *wound himself* in the act of perpetrating a deeper spiritual insight.
The act of painting becomes a recreation of earlier rites, old ritual wound-
ings, the deep scars of initiatory memory. He sees his work as an extension
of his intiatory life, the process by which he is transformed from a youth
into a fully fledged adult. The work of art becomes an icon of himself
which can be viewed in his own absence—that is, when he has returned to
the realm of the Dreaming at death.

The power of the painting must be seen for what it is: an icon depicting the man, his knowledge, his suffering and anguish, almost as if the painting were the man's double. Painting his Dreaming must be viewed as a celebratory act, a catharsis, a moment when his entire being is solemnized on canvas. Perhaps it is hard to attribute such feelings to a traditional person, a paleolith, someone not embued with the modern condition of *angst*. But we must remember that Aborigines see themselves as custodians of an ancient tradition, one which reaches further back in time than we can imagine. They are conscious of the fine line they tread between observing the rich heritage of the Dreaming, and succumbing to the white man's disease of materialism. They are aware that they possess nothing but their stories, their rituals, and their traditions as part of a timeless inheritance. They are also conscious that white men possess things, toys, objects which do little more than possess them for a short period of time. It is what Aborigines are afraid of most: of being possessed by artifacts which are not of their own making, and which are also steeped *in* time.

Aborigines are fully aware of the savagery that permeates their nature. Their strong sense of morality, of what is right, is governed by a need to survive. Aboriginal ethics are linked to the most basic imperatives. Hunger, freedom, good kin relations, right knowledge, the ability to invoke the Dreaming heroes, all are important to the resolution of tribal difficulties. When a man is speared for committing a wrong-doing he understands the purgative nature of blood-letting, of subjecting himself to a formal spearing. The wound he receives is tantamount to a badge; it announces to all that he has been subjected to a savage encounter which cleanses him of his guilt. Not to be speared, and so to be hauled off by white authorities to face charges in common law, removes him from the fray. Often against his will he will have been taken away to face a charge that can only be truly ameliorated by the shredding of blood. Jail, for him, the white man's penitentiary, is not a *wound* as such; it is a living hell which precludes any possibility of a pure act of savagery being inflicted upon him. A man wants to feel pain in the face of his tribespeople's displeasure. He wants to see his vital essence, his blood, being shed as propitiation.

It is this moral imperative that influences their art. One often senses that a painting depicts in stylized form the incisions made upon a young man's body during his initiation. One sometimes feels the pain of circumcision, the splitting of the penis into two halves of flesh (sub-incision), or the rubbing of charcoal into the open wounds on his chest as one gazes at a canvas. The painting becomes the actualization of his pain. It becomes the expression of a desire to endure physical hurt in the interest of growth. One

begins to realize that Aboriginal art is more a manifestation of a transformative process than it is the depiction of primary objects.

The Aboriginal artist brings these realities to bear while he paints.
When he lays his canvas on the ground, often in the full glare of the sun, he
knows that his work is about to be brought down to earth. As the Walbiri
say, "Walbiri, we live on the ground." A man must be near the ground
before he can feel its suscitations. He longs to be stirred up, to be brought
to a pitch of inner excitement by the memory of his Dreaming country, of
the great ceremonies he has participated in during his lifetime. Above all he
wants to engage in an interior dialogue with what pre-exists in his own
mind. Such a mode of understanding enables him to come to terms with a
deeper knowledge of the mysteries of the Dreaming. He becomes an example
of Aristotle's dictum, that of identifying his soul with what he knows; or
with Aquinas's belief that 'knowledge comes about in so far as the object
known is within the knower' (*Summa Theologica* I,Q. 59, A.2). His aesthetic
experience is thus derived from an innate knowledge of absolute values in
exaltation of the pure consciousness of the Dreaming.

It is at this point where savagery and sanctity merge. The unleashing of
powerful psycho-spiritual forces brings about a deep transformation. The
Sahitya Darpana (III, 2-3) regards such an intuitive bout of ecstasy as a blind-
ing flash of light of transmundane origin, impossible to analyze, and yet in
the image of our very being. The knowledge of such ideal beauty cannot be
acquired, but is born within us. It comes about through the rectification of
the whole personality, accomplished in a previous condition of being. For the
Aboriginal artist, such a rectification would have come about through
knowledge of this totem, not by any conscious attempt to modify his behavior. He would have entered into a state of "deep sleep"—that is, he would
have attained to a condensed understanding of the Dreaming through an act
of meditation, or indeed one of ecstasy. He would have come to a deeper
knowledge of the Dreaming through a cessation of the consciousness of particulars. Nothing exists for an Aboriginal artist but the possibility of rendering his visionary landscape in all its wholeness.

On many occasions I have sat beside Aboriginal artists while they work.
I am always struck by their concentration, by their unwillingness to break
off what they are doing in order to engage in conversation. If they do decide
to talk, then it is not to talk about ordinary things. Wimmitji Tjpangarti,
an old Pintubi from the Balgo Hills whom I know, will never talk about
any other subject other than his Dreaming, for this is his *only* reality. With
his voice and his hands, he will travel across his visionary landscape, caress-
ing its contours into life. At no time does he open his eyes while he speaks.

These are always closed even when he paints. He has all but dispensed with sight as a mode of reference. Wimmitji relies on touch, on the tactile presence of form, on where his inner landscape is transporting him. What is occurring, if without any obvious signs that I can see, is the man undergoing a prolonged state of recollection, or regathering his memories of the past. He has allowed himself to be raised to a level of pure intelligibility whereby all that he sees, all that he feels, and all that he knows is rendered as image on canvas.

Wimmitji is a different type of artist than most. He is not a "painter" in the modern sense. He does not set out to personalize his art in any conscious way (although inevitably his style is himself). He is not conscious even of painting a "work of art." Nor does he conform to any criteria of what an artist should be. If he is inspired, then he is receiving his vision from a suprapersonal source, not one governed by the here-and-now. At no time does one feel that he is engaging in an artistic activity when he crouches by a canvas, holding one trembling hand with another as he endeavors to place dots on the tableau. What he is doing belongs to another order of expression entirely. For he is immersed in what Ruysbroeck called his "incomprehensible nobility and sublimity." Wimmitji, like all truly great savage artists, is exploring the very ideal of himself as it manifests itself in the images that he paints.

Of all the artists currently working in Aboriginal Australia, perhaps those residing at Wirrimanu (Balgo Hills) in far northwest Australia are closest to the spirit of savage art. Many of the older painters there are people in their middle sixties or even older. Their early life was spent in the Gibson and Great Sandy Deserts before their encounter with white men. Only in their mature years did they come to live permanently at Balgo, thus relinquishing their traditional nomadic ways. Such artists bring to their work a unique perspective, one which is the preserve of a dwindling group of people throughout the world. They still regard their land, and the totemic wanderings of the Dreaming heroes, as a living reality. To speak with these people, and to observe their paintings *in situ* is to encounter a powerful sense that here lies a prototypal art environment. Here, so to speak, are the caves at Lascaux and Altamira. Here are the great rock frescoes of the Tibesti Mountains in central Africa. The Balgo artists portray a spiritual dimension which has all but disappeared from the world.

Savage art has the power to render us new again. Once we have accepted its refusal to allow its message to be subsumed by aesthetical values, we begin to know what it is like to confront the numinous in a work of art. We are confronting all those non-rational elements associated with "daemonic dread" which we like to think have passed from our spiritual repertoire for good. We are coming to terms with all those daunting aspects of the numen which once, in a previous life, allowed us to experience feelings of positive self-surrender to a force greater than ourselves. Savage art, and Balgo art in particular, is a shamanic art whereby one begins to identify with magical transformations, and with an indwelling and often dormant feeling of exaltation. It is at this point that a very real sense of *mysterium* can be experienced as something essentially positive, noble, and specific in character, something that can bestow upon each one of us a beatitude beyond compare.

Essential also to the unique power of this art is the knowledge that every painter is a "holy" man (or woman). As an artist he or she is no longer a mere person, but someone who partakes of the wonder and mystery of the Dreaming. At once the artist is seen to belong to a higher order of individuals, the people of knowledge, the *mekigar* or sorcerers. Artists are, in a sense, "possessed" by the archetypes of the Dreaming which they are powerless to dismiss. The very fact that they cannot clearly identify the special character of the Dreaming, except by way of visual or ritual metaphors, makes it a supreme cultural exemplar. The Dreaming settles over these artists in the same way as the spell of Niniane settled over Merlin in the forest of Broceliande. Artists become a part of its primeval motif, and are thus overpowered by its timeless link with the source of all creation. In so doing they find themselves transformed by a nostalgic worship of dissolution and a loving sense for the mysterious descent into the womb of nature.

Here lies the fascination of savage art. It attracts us because of the compelling force of its argument. No other art is more capable of neutralizing the idea of a cosmos in turmoil. It permits us to attain once more a feeling of the sublime after an unremitting pulverization of our senses. In its depiction of the Dreaming, savage art allows the viewer to pass through all the filters and foundations of existing matter. As a visual language it becomes a form of incantation which makes it possible to return to nature. We *see* the cataclysm as if emerging from the Dreaming, calling upon us to rediscover life. Life's intensity becomes less remote, less given over to our age-old desire to intervene in the formulation of the divine. In this way savage art orders us to lose ourselves in the imagined form of the Dreaming and to be impelled by its force.

It is through the skin that metaphysics must be made to re-enter our minds. Savage art penetrates that layer and forces us to come to terms with what we have denied for too long—our unwillingness to see destiny as anything more than a play of individual consciousness against the blind vigor of existence. For savage artists such a view is too limiting. They are primarily engaged in transcending such an experience by invoking a loss of selfhood with the aid of one's totem. In the end they become something that they have always been: a work of art themselves, although fashioned in this instance by the imperious power of the totem as an agent of growth.

Savage art demands much more of us than we are often prepared to give. Because we are inured to the ritual demands made by this form of art, we see it as the product of a so-called "primitive" mentality. The artist is depicting pre-civilizational images. He is asking us to shed our notions of what constitutes being civilized. He wants us to enter into a pre-conscious phase wherein the primary emotions dominate the way we encounter reality once more. Fear, pain, sexual potency, anger—all such emotions seem to dominate savage art at the expense of those more sublime feelings we associate with refinement and sensibility. It may be argued, however, that it is these feelings which have so deranged modern art because they have separated the mimetic act from its origins. The modern artist has lost the ability to call upon those instincts out of which intuition grows; instincts which a savage artist relies upon to invoke the imaginal world. The difference between oneself and the modern counterpart is that the Aborigine has not attempted to humanize what is seen. Instead there is a reliance on a certain psycho-spiritual condition to govern one's actions: intoxication. It is the intoxication that follows in the footsteps of all great desires, all strong emotions; the intoxication of feasting, of contest, of the brave deed, of all extreme agitation. Its essence lies in the feeling of plenitude and increased energy that flows from it.

Intoxication compels savage artists to *idealize*. This is not to say that they subtract or deduct all that appears petty or secondary; rather they express a tremendous *expulsion* of the principal features in an attempt to give form to their own feeling of abundance. What they see and what they desire is overladen with energy, and they need to appease this abundance. In so doing they create a work which partakes of something that is less sublime than it is energizing. Savage art provokes us to want to regain contact with our primary emotions, emotions which have remained suppressed in the interest of refinement. What the savage artists ask of us is to explore anew that wellspring of feeling which encourages conception and birth (potency), those primary integers of renewal and world-creation. Indeed, the continuing existence of the great Dreaming heroes among Aborigines today are a result of this super-abundance.

—22—
Tribalism

Avram Davis

Given the current depth . . . of the problems raised by technological develop-ment, no political action in the normal, strict sense of the term is adequate today . . . we can always issue decrees and pass laws, which sufficed for the problems of society one hundred years ago. . . . This signifies that during the five hundred thousand years of our existence, we have developed in a specific direction, and now we are suddenly being asked to change . . . [but] we cannot suppress half a million years of evolution in a few short years. What we can predict for sure is that if technological growth continues, there will also be a growth of chaos.[1]

Hollywood-style materialism, rampant technology, and growth-oriented ways of thinking now enjoy world-wide popularity. America is desperate to increase its growth and China wants to put 200 *million* cars on the road within ten years. Everyone wants a taste of the good life, and the good life is defined by Madison Avenue.

One of the wonders of modernity (and post-modernity) is its ability to grind up everything it touches into a homogenized blend. It is a blend (we often call it "culture") of anonymity and consumerism. Herbert Schneidau calls this process "westernization" and believes it will be the fate of all peoples. "[People] take their place in a long procession, some of whom came to harass and remained [to be transformed]." The ideologies of East and West make no difference. Marxism and capitalism play upon the surface of cultural change—they bend to the wind identically. "Whether the [westernization] of the peoples is done by Capitalists or Communists, doesn't matter."[2]

But there *are* communities that embody Thomas Foster's definition of ideal models for the post-modern age. They are spiritually alert, opposed to waste and consumerism, and are ecologically conscious.[3] These societies are, to a greater or lesser degree, *tribal*; able to mitigate or deflect the more obnox-ious elements of modernity and technology.

Modernity has traditionally seen culture as progressing from the primitive to the modern. Thus it has always placed itself at the pinnacle of worth. Only in this way has it felt comfortable with the enormous destruction and co-option of the more "primitive" cultures it has undertaken. It has left us, as Peter Berger puts it, "homeless." Western, and increasingly Eastern, culture is without "home."[4] And such a loss includes identity, minds—the very essence of our lives, thus presaging one of the most serious sadnesses in the world: the vanishing of ethnic/tribal groups.[5] This trend is often defined in terms of "The West" swallowing up the world's diversity, creating a vast Coca-Cola wasteland.

Yet, here in America, the heartland of the "West," things are awry. Americans yearn for community. It can be argued that the need for intimacy, community, and kinship grows stronger as the alienating power of modernity insinuates itself in peoples lives. Kinship is one of the primary ties that bind. This is why family, in whatever form it takes, is valued and protected. Kinship is the cornerstone of the tribal vision.

When free to choose, most people prefer to live in manageable societies that are controlled and motivated by a web or responsibilities, love, obligations, and personal contract. When we look we see relations of an intimate kind; something more than mere mutual economic gluttony. It is a relationship that is very personal, very intimate in its essence. "We think of ourselves not as human beings first, but as sons and daughters . . . tribesmen, and neighbors. It is this dense web of relationship and the meanings which they give to life which satisfies the needs [that] really matter to us."[6]

Tribalism is generally viewed as regressive, primitive, and archaic. In Western culture it is something to be progressed *past*. To be tribal is to be archaic. Besides being anachronistic, there is the fear tribalism will lead to conformity, lack of autonomy, and the subservience of the individual to the needs of the community. What is the information gleaned from traditional culture worth? What benefit can we (moderns) derive from tribalist/particularist cultures?

The heart of the *particularistic* and the *tribal* consists of the close and the intimate. It is exclusivist, insisting there is an "outside" and an "inside"; a "belonging" and a "not-belonging." This should not be construed or translated as racist. Rather, it is the particular and the familiar. It rejects rhetoricians who speak of "healing-the-planet" when they are without any specific responsibility to family, group, or land. The particularistic is allergic to ad and fashion. The tribalist is dictated to by the past and the wisdom of tradition. It defines and enters a future with its feet planted firmly on the wisdom of the past. It maintains that neither planet or country consists of one universal people with a shared vision but rather consists of many thousands of separate peoples.

The heart of the universal consists in the ideological. It embraces the large, all encompassing theory. It abounds with grand visions and sweeping testimonials: mass conversions, mass movements, mass fads, and fashions. It is a classically "American" system, but curiously it is also Marxist. For it promises much later, if we but forget our local affiliations now. It is a dream that must ultimately say to the Lakota (Sioux), the Jew, the Amish, and so on—"Your diversity is a romance and perhaps a menace. It is acceptable only to the degree it is powerless. After that there are two choices: Wounded Knee or friendly assimilation."

The strength of the tribalist vision resides in three major areas. In the notion of a specific region, by which I mean a specific consecrated acreage. In the notion of kinship which itself has many subthemes. And in the notion of a common past which moves toward a common future.

The nature of land—our relation to it, is of primary importance to the tribalist. Between tribal groups the roots that first caused attachment to land differ. There is always a strong personal bond to a particular land base. This is what warrants particular emphasis. Affection and attachment to land have been much discussed—Paul Shepard, Erich Isaac, Yi Fu Tuan, Mircea Eliade are just a few of the many writers who have made this one of the main themes of their work.

Land may be found in the particulars of a route that punctuates and gives meaning to a childhood memory. Land is the grammar of a neighborhood. The nature of land may be found in the very specific and unerring knowledge of path, burse, vectors, animal concentrations; the very smells of the land reflected topographically. It may be a topophilia of love or it may be something much more prosaic.

Sometimes land is thought of as a thing. As acreage. A commodity to be bought and sold. Sometimes it is seen as an idea, a construct imagined and defined according to one's political, ethnic, religious, or economic predilections.

But the repercussions of all these attitudes are political. They are political in how they help us define and live our lives. For attitude prefigures action. As land is perceived, so it will be treated. If land is seen by the person who lives on it as something sacred, all subsequent relationship to it will differ from the person who views it as merely something to buy and sell.

As mentioned earlier, one of the three seminal issues in the construction of a tribalist mindset concerns land. Not in an "Earth First!" or environmentalist sort of way (for I see these organizations as part of a larger, *universalistic* mind-set), but as an attitude that is key to tribalist self-definition. The tribalist does not nurture or protect his land because it is *politically correct*, but because it is part of religion itself—it is part of the fabric of his internal self definition. This is very different in kind from the loyalty inspired by the *merely* political ideology of environmentalism.

Tradition weaves connections between a people and the land. These connections may be applicable on every level of tribalist thinking, whether Jewish or American Indian. They are all grounded, as it were, in the *sacred*. That is, the relationship is rooted in sacredness. Sacred duty; sacred covenant; sacred notion of ethnicity and self.

Generally speaking, the tribalist vision views land not only as sacred in itself, but as a conduit between the greater whole and each individual. Sacredness is *passed* between land and human.

Westerners are often divorced from spatiality and the concreteness provided by a specific grounded past. Their pasts do not cry out to them. Most Americans do not experience the land as a trust or covenant. Is it any wonder, then, that this ownerless, stewardless America seems increasingly without hope, without nurture, without future as it fragments into ever greater anger and chaos? Without a firm grounding in land connection no people can sustain itself. Like the myth of the son of Earth, Anteus, a culture remains strong so long as it is connected to a specific geography. When this is removed, it is left struggling in thin air and will expire.

To re-embrace the tribal is to rekindle particular affection for one's own land. Rooted in a love that is frequently described in the most intimate of terms, such connection has served as a shield against endless theorizing and abstraction. For example, after the Jews went into exile, the ultimate reward for living a holy life and for fulfilling all the laws and commandments given over by the tradition was to be able to eventually live, die, and be buried in the land.[7]

The notion of centrality, of a geography rooted in past and divinity, has lent strength to the tribal mind from generation to generation. For the Jews, as for most indigenous Americans, the "holiness" of the land is derivative. Its holiness is derived from the fact that *Godness* saturates it. Its holiness flows through it. Because holiness flows through the physical membrane of the land *this* is what gives it its holiness. If this flow is broken, a loss of totality and wholeness is experienced.

Not only humanity will suffer but the land itself. It is written in Isaiah, "The earth mourns and withers, the world languishes and withers; the heavens languish together with the earth. The earth lies polluted under its inhabitants, for they have transgressed the laws . . . broken the everlasting covenant."[8] The damage is done by human beings breaking faith, surrounding the covenant made between themselves and heaven and between themselves and their ancestors. And it is *felt* on many levels; human and non-human.

The second notion that supports the tribal vision is the assumption of some sort of kinship relationship between members of the tribe: the primacy of relationship itself, the notion of reciprocity, and the empowerment of intimacy as a redemptive force.

The current evolution of the political debate between varying ethnic groups within the United States necessitates a re-examination of various definitions of ethnic relationship and identity. It is a crucial issue if we are to understand the healing possibilities of kinship. Jews and Indians have always been known as "people," "race," "nation," or "tribe." All of these words have much to teach us about the nature of ethnic redemption.

Togetherness is an essential element in the kinship bond among tribalists. It may well be the most *fundamental* element of a tribal polity—that is, the understanding within the group that they are, in whatever form or fashion, related.

For example, there were once twelve tribes of the Jews. But these tribes were destroyed and their remnants collapsed (as it were) into one tribe—the tribe of Israel, also called "Am Yisroel"—the nation/people Israel. But the term is slippery. For when we say "nation," whether in an Indian or Jewish context, we say something quite different from "homeland." They involve quite different sets of responsibilities, obligations, hopes, and fears. The term *Yisroel* (Israel), for example, refers to a nation state located in the Middle East but also to the Jews (as a united, single group) dwelling in the world. It is clear from all statistics that Jews view themselves as "bound" together by more than mere religion. Especially as most Jews are not particularly religious! Similarly, Jews who inhabit the homeland of Israel see themselves bound together with all other Jews and the Land as the physical focus for that union.

Ethnic specificity has long been an imperative for tribalists. For Jews it dates back to the very strict admonitions about intermarrying with the inhabitants of Canaan (the neighbors of the biblical Jews) as well as the vocabulary by which the Jews are described throughout the Torah: "Children of Israel," "Seed of Israel," "Tribes of Israel," and so on. All of these may be seen as codes (albeit not very subtle ones), describing an eponymous ancestor from whom all are essentially descended.

John Berget points out that the traditional tribalist is quite different in certain fundamental perceptions and analyses from a non-tribalist. The tribalist experiences a "closeness to what is unpredictable, invisible, uncontrollable, and cyclic. . . . [They do] not believe that Progress is pushing back the frontiers of the unknown, because they do not accept the strategic diagram implied by such a statement. . . . The unknown is constant and central: knowledge surrounds it but will never eliminate it."[9]

Kinship and all that is implied in it—intimacy, reciprocity, relationship, and so on—are vocabularies that may be woven into our larger American polity with but little change. In the deepest way, "relationships . . . can be extended fictively to people who are not related."[10] This may be understood in two ways. Firstly, people are brought into a *family* and secondly, outsiders see how it is done and begin setting up similar structures in their own circles. Tribe, in spite of being a familial, blood-based grouping, can include members who are "converted" to the tribe. Deep community demands relationship.

Religion may often be tangential to our lives, but the need for a sense of "belonging" is firmly fixed.[11] Jews use their synagogues and community buildings extensively, but this reflects a use of *memory* rather than religion. Of course there are also a sizable number of "religious" Jews but, the matter of memory is crucial. It is a perception closely shared by all tribal peoples. Indians, as Russell Means points out, are " . . . still in touch . . . [with] the prophesies, the traditions of our ancestors. We learn from the elders . . . ."[12]

Memory is the forge in which tribal people recall themselves and perpetually recreate themselves. It is the memory of a common past, the acknowledgment of a shared present, and the assumption of a common future that lends the tribalists their foremost continuity. Memory is the realization of religion for man but it is also the simple recognition of lineage. This realization is inculcated into tribalist peoples at every level of their development.

A charming *aggada* (an illustrative story) that begins a chapter dealing with the Jewish people in a children's primer illustrates this.

> . . . we must put the needs of Klal Yisroel [the entirety of the Jewish People] ahead of our own sense of importance. According to one story, when God told Moses he would die before the people entered Eretz Yisrael, Moses did not complain to God. . . . Instead he said . . . Please choose a leader to take my place, so the people will not be like a flock without a shepherd.[13]

Another children's book is even more explicit about this notion of responsibility and continuity.

> [All these empires vanished], the Hittites, Canaanites, Phoenicians, Assyrians, Babylonians, Persians, the ancient Egyptians, the ancient Greeks, the ancient Romans. [But there] still exists one people . . . [with] the same ethnic unity it had 4,000 years ago when it started out in history. This people is still as mentally alert and alive today as it was then . . . the Jews.[14]

This notion of past prefigures a common sense of future. A future that is inhabited by one's kinsmen. Russell Means refers to this when he expounds on a shared past of Indian people which will lead inevitably to a common shared future. A future that is assumed because of tribalism's shared memory. "And when the catastrophe is over . . . American Indian peoples will still be here to inhabit the hemisphere."[15]

The notion of continuity, both within the religio-spiritual tradition as well as ethnic self-awareness, is epitomized through the notion of transmission. Transmission of the codes, the habits, the rituals, the blood lines.

One of the main codes for Jews (the Mishna) recognizes this. When the Mishna describes its own creation and how it transmits from ancient times into the hands of modernity, it presents its own genealogy as an utter reliance on memoried transmission.

> Moses received Torah from Sinai and he handed it on to Joshua
> and Joshua to the elders and the elders to the prophets and the
> prophets handed it on to the men of the great synagogue. . . . [16]

If one of these generations is broken, the whole is lost. But because none of them *were* broken, the whole is carried on to the next generation. Indeed, it is possible that of all the various necessities that undergird the survival of the Jews and other tribalist people, this reliance on memory and sense of common past and common future is most enduring.

For Indians, moving from a tight-knit reservation world to an urban setting is unnerving because of the loss of a memoried community. The common sense of peoplehood and a commonly shared past leading to a future is what is destroyed in the vast anonymity of today. Peter Snyder's study of the Navajo points to a high level of self-ghettoization (or as he calls it, an "enclave"), even when there are only small numbers of urban Indians.[17] And this is because their peoplehood cannot be maintained apart.

The tribalist has been described as stupid or backward; a roadblock to progress; cunning and idle.[18] But much of this complaint is rooted in a misapprehension. The tribalist is defined and commanded as much by the wisdom of tradition as the allure of future progress. The tribalist walks on roads beaten flat by his ancestors. In a manner of speaking, the tribalist becomes, perpetually, his own history. He is shaped and defined by the past and therefore must live always partly in the past, even while he must make and remake the future. His concern is less with what makes him personally grow, than with what is needed by his people, his tradition, and his soul. For his soul is connected to the whole. It is a very different area of cultivation than ego.

No one aspect of tribalness necessarily offers protection. The accumulated lore (or wisdom) of any number of tribal peoples has not been sufficient to save them from extinction. Memory can be extinguished or manipulated—this is after all the era of advertising and propaganda. This is Milan Kundera's point in his *Book of Laughter and Forgetting* when he depicts bureaucrats airbrushing various political figures in and out of existence as they rise and fall from favor.

Americans are bound together. But it is an increasingly nominal bond, consisting of certain, specific laws and economic premises. It is defined by clichés: "the American way of life" or "the American dream" which continue to hold an enormous power for much of the country and much of the world. But this dream has become a nightmare for many. It is rapidly eroding even the cliché benefits that drew so many.

There is a limit to the degree to which a society can be psychologically dysfunctional and still function politically. It is not a question of the quality of life in terms of material wealth, but rather a question of inner peace—the quality that earlier generations called wisdom, that is increasingly lacking in these times. The exploitation of all that was sacred to the past for the sake of financial gain or political advancement is a complaint much bandied about. But the real truth of the matter is the biblical expression that one "who sows the wind will reap the whirlwind." There is an uneasy feeling that the whirlwind is blowing just outside the door. Increasing numbers of Americans feel its chilly first blasts. Sociologists and theologians both point to a breakdown of the dream that holds us together.

> What has for a long time been dismissed as idealism seems to be
> the only realism possible today.

The tribalist knows, in his heart, that everything that comes from the land is a blessing—even if it kills him. The *wholeness* of it is a blessing. "[This is a] land which God, your God, cared for; the eyes of God are always on it, from the beginning of the year to the end of the year."[19] And this teaching cries out to us for greater understanding in our modern lives.

A religious reflection of this was summed up by Rabbi Tarfon: "The day is short and the work is great and the laborers are sluggish and the wages are high and the householder is urgent. . . . The work is not upon you to finish, but nor are you free to desist from it. . . ."[20] Everything is bound by bonds of intimacy. Humanity is responsible; a charge given over from the highest heavens. We are responsible on every plateau of meaning. We are consecrated by our relationship to land, to each other, and to every bit of life uniquely put upon the planet.

A tribe lives or dies by its attachment to a particular land base. It either holds it or retains the memory of its attachment with an attendant yearning to return. When both of these disappear, the tribe will disappear. The relationship between a tribe and its land is generally a bond in perpetuity, first given to ancestors and handed down as a trust to successive generations. "That your life and the life of your children may be prolonged in the land, which the Lord promised He would give to your fathers as long as the sky remains over the earth." This sentiment is one which we moderns need to embrace. We will live or die by our attachment to the earth and its creation.

Attachment is interwovenness. Attachment is intimacy. Intimacy, in tribal terms, reflects reciprocity. If the notion of land is fundamental to the sociological landscape of tribal self, then the glue that holds this together is reciprocity and mutuality.

In such a schema the living world is a single entity. It is connected and perceived as being somehow all of the same cloth. Human and animal, animal and plant, mountain and humanity. All share the same fate; all share the same genesis. "For that which befalls the sons of men befalls the animals, even one thing befalls them; as the one dies, so dies the other, yes, they have all one breath, so that humanity has no pre-eminence above an animal."[21]

The process of leavening up human and non-human, animal and nature; of perceiving human and landscape on a plane that is equitable (though *not* necessarily constant), is a firm part of the tribal. All creation is bound together. Even when it is not understandable—even when the creation is annoying (to human senses). "Rabbi Yehuda said, in the name of Rav: everything the Lord created in His world has a purpose, even the things that you may consider to be unnecessary; such as flies, fleas, and mosquitoes, are part of the creation."[22]

So fundamental is this central idea that it even transcends the attitude to specific land loyalty. The idea of reciprocity must pervade tribal lifestyle in order for existence to continue, especially in the harsh present day. "The whole world, man, animals, and birds, all find their food in what was created in heaven and on earth. All the residents of the world are governed by one and the same star."[23] This common fate and boundedness together makes a situation of all life in partnership. All are related, members of the same biological tribe.

Though human beings have the ability, even sometimes the "right" to destroy and to take, this is not the process that supports the universe. This is not the way in which Heaven set up the basic structure. The Jews learn "See, how all the Lord's creatures borrow from one another?"[24]

The emphasis is upon *borrow*, rather than take. The nature of reality is infinite "borrowing." Each creature borrows what it needs in order to survive. Conversely, careless *taking* leads to destruction.

But again, arcing over this relationship is the idea of intimacy. That is, the *Infinite* cares about the *finite*, and is connected. This aspect of intimacy, of the Infinite as father or mother—is a reflection of tribal mind and for all tribalists it defines the relation between people and land as well as the relationship between the Holy One and humanity.

> [The various aspects of the landscape] are seen to be brothers or
> relatives (and in tribal systems relationship is central), all are . . .
> necessary parts of an ordered, balanced and living whole.[25]

It is not enough to say that everything possesses a life. This lends itself to the notion that everything is a sort of protoplasm, directionless, formless, and without being. No, most tribalists conceive everything as alive, as being part of a great drama; but also as having direction and being directed. And direction is the Great Spirit. And this spirit is caring; caring of every creature, every stone, every word spoken or fear expressed.

Nothing goes on in the world, much less the special ancestral Land, that is outside the ken of relationship. Every animal in travail is heard by the Spirit of the place—by God. Nothing is too small to escape the notice of directing Force. And this notice, for the Jews at least, in its essence, is benign. "The body of the gazelle is slender and she has trouble giving birth. What does the Infinite One do? He calls up a snake which bites the gazelle enough to relax her entire body. What does the Infinite One do after the gazelle gives birth? He brings her an herb which she eats, this helps her to recover from the birth."[26]

The principle of intimacy and reciprocity exists throughout the world and is prominent in the cosmology of most particularistic peoples. Still, the notion of reciprocity may be encapsulated in a smaller word—the *local*. For the *local* is where holiness is most manifest to the tribe. It is in the *local* that the tribe experiences the transcendent most profoundly, but also partakes of the intimate interchange between heaven and earth. It is within the *local* that the Infinite becomes (as it were) tangible. For the local is Heaven's special, specific gift.

The world cries out to us. But respond. . . . Where? How? I believe ultimately it is in the intimacy of tribalism that mercy will lodge and redemption be found; and further, it is in the honoring of tradition that justice will, in the years ahead, find a safe heaven.

—23—
The Sacred Womb

Georgianne Cowan

At conception another presence nestles alongside one's soul. The physical embodiment is not yet perceptible, but there is an internal quickening that evokes a promise of life unfolding. The uterus's pear-shaped sea of fertile juices protects the embryo, and the membranous material of the placenta nourishes the small life as it incubates within.

The recipe for life is drawn from the pharmacopoeia of the body's cellular memory. In includes DNA, a smattering of neutrinos, brain tissue from extinct reptiles and the chromosomal blueprint of this primitive zygote's destiny. A creative explosion takes place; in a matter of milliseconds after the sperm has touched the egg, the impetus towards life takes over. There is nothing to control here, nothing really to be done, no way to force this process. The transformative weaving of tissue to bone and water to blood coalesces only at the proper moment—when life's essential ingredients have been intermingled and gestated inside the warm fires of the womb. An inherent law of nature is at work; growth cannot be rushed, nature will unfold organically in the time span deemed necessary for survival. This numinous mystery is propagated by the realm of fecundity, where control is obsolete and nature is the poetess of life.

The impetuous seed of creation does not exactly come forth on little cat feet. Rather, it takes hold without deliberation. A woman's womb, once virginal except for the cleansing rivers of menstrual blood flowing through her, becomes possessed by this amazing force of life. A new sensation flutters inside, a veritable stranger takes up residence inside its mother's belly. The queen has allowed one of two million insistent sperm into her domain, and there is no turning back as cells divide and the royal egg transforms from zygote to blastocyst. The force of life is aggressive and steadfast as the embryo recapitulates the stages of evolution. The throes of this alchemy unfurl a sensitive sea-like creature who develops a tail, fins, and eventually tendrils of arms and legs. For 266 days inside the womb, the mysteries of the life-cycle are enacted as the initial protoplasm expands, forms, and blossoms.

Who is this new visitor? What is this oscillating, fluid mystery inhabiting her body? Is it human or vegetable, monster or something completely unknown to our collective psyches? The *stranger* takes on a somewhat ominous quality, simply by the nature of its invisibility. Who is to know what is forming inside? The answer may rest in our collective knowing, on the faith that our species gives birth to its own and will eventually birth a human being. Is it not unlike our faith in the return of spring after a winter of darkness? In the velvet womb of the earth, in the dark womb of our soul, we've peered into the shadowed void of the unknown and trusted that from fallow fields of dormancy will spring new life.

The Myth of Demeter and Persephone personifies this cycle of transformation by combining human, mythic, and natural cycles into one parable. Demeter laments the loss of her daughter Persephone to the underworld as the dead of winter chills the landscape, and rejoices with new life at her return in the spring. The cycle of life, death, and rebirth is enacted in the womb of the earth and the psyche of humanity. In another more condensed form of cyclic transformation, our daily ritual of sleep surrenders us to the dark underworld of the night—we sleep, we dream, and awaken again and again to the rebirth of light—a symbolic beacon leading us to return to life each day. We understand from this that all creative manifestation requires patience, faith, and a descent into the darkness of the unknown; from the evolution of egg to newborn, winter to spring, and waking from a night's eternity of dreams.

The dark womb is a vessel for containing the rawness of physicality, a place where blood, tissue, and bone alchemically transform. The fertilized egg sets the directive and millions of cells abide with the subtle needs of this highly physicalized moment of creation. All the ingredients are generously provided to sustain this fragile life, in the same way the bounty of the earth nourishes us outside the womb. The whisper of a benign new spirit breathes its presence through the suddenly receptive cells of the body. A window opens, holy winds impart the inspiration of life like the random casting of seeds onto the soil. This metamorphosis of matter is intimately related to the invisible stirrings of the spirit where, in this instance, a divine marriage between heaven and earth has been consummated in the form of a burgeoning new life. The medieval alchemists in their quest for "gold" likened the vessel where they performed their alchemy to a second womb where opposites—male, female; heaven, earth—merged and a latent bond was brought to light. These acts of chemistry were intended to bring forth the "gold" that dwelt implicitly in all matter. Such experiments were an enactment of the transformative alchemy embodied in a woman's womb.

It takes ten days for the egg to travel through the fallopian tube and to implant itself in the lining of the uterus. It is the domain of the mother to incubate this life and to become one with the process. She is the sacred container. Time, now more than ever, follows a creative continuum. Each instant unfolds a sprouting of life, from the buds of small fingers to the weaving of the spinal cord. The bodily wisdom of the feminine cellularly understands this creative principle of existence, and the circular nature of time. It is in the fertile womb that immanence has true meaning—where the relationship of self to source is an original blessing—a compass for locating home.

If we journey back to our prehistoric origins, we see that the primitive forms of life that inhabited the ancient seas 4.5 billion years ago hold essential secrets to our ancestry. All life on earth evolved from these simple protozoa. In time, the protozoa evolved into sea creatures, then much later emerged from the waters and slithered onto land. Our reptilian ancestors came forward dripping with the memories of swimming in the amniotic tide of the sea. We discover now that the cells in our bodies are composed of the same foundational substance that floated in these primal seas, and the salinity of the ocean matches that of the amniotic fluid rippling inside the womb.

Our dreams sometimes evoke a deep connection to this timeless, visceral world; to the one-celled organism, to the slimy brine of the sea, and to the bloodlines of our ancestors. . .

I dream of a pot of boiling water with a swan-like bird carcass floating in it and also a sea creature like a sea anemone or shelled creature combined. The sea creature is alive and moving. The bird is part meat and bones—large like a swan. I reach into the pot and pull up the carcass and the collar bone and skull and three long bones emerge out of the water and align themselves like a sacred Talisman, part cross, part ancient bird.

Essentially, growth evolves from the egg coalescing into a more complex conglomerate of cells and layer upon layer meshing into form. Physical density is added through the incorporation of earth, water, fire, and air and the mysterious magic of life's presence . . . or spirit. The baby's growth is constant as it buoyantly floats inside the cavernous walls of the womb. After ten lunar months, the cosmic time clock gongs inside the pelvic bowl, and the journey of pushing, pulsing, and parting is begun through the birth canal. Baby descends head first, through the arch of mother's thighs, toward the earth. This fragile, wet, amphibious-like creature is made of the same cell tissue and blood of her/his parent's hearts, not to mention the same substance that nourishes and holds us through eternity—the regenerative web of what we call nature.

Pregnancy is a time of connecting deeply to a creative source of being. This is the one time in a woman's life when she can feel the development of another soul inside her own body and a time when the man as protector of life can witness a part of his essence growing inside his mate. In this moment, we have an opportunity to grasp our integral role in weaving the fabric of life, all the way back to our ancestral family of sea anemones up to our present moment as humans on the earth. This pregnant incubation is the quintessence of our familial relationship to nature and the universe. The miraculous unfolding of life during pregnancy is one of the most profound and mysterious gifts of existence.

The energy of creation, in its quest for nascency, moves in a tremulous, creative motion, aligned with a cosmic rhythm that is akin to that of the earth's. Nature's principles of growth and generation are analogous to the life-giving matrix of a mother's womb. In ancient Greece, for example, the *omphalos* (also the *umbilicus*), or center of the earth, was considered a place of reverence. Delphi (derived from the Greek *Delphys* which means womb) was one of the great centers of antiquity and the temple at Delphi was constructed to sanctify the place which was said to be the "Temple of the Womb." It began as an earthly temple for Ge, the earth mother-sustainer of life and was the consecrated womb center of the earth, where life was said to have been conceived.

Countless other ancient cultures endowed the earth's womb with this same supreme meaning: it was called the *beth-el* by the Hebrews which meant the "dwelling place of deity." In Mexico at the site of the ancient civilization of Teotihuacan, two pyramids were erected, known to be tributes to a nature goddess and storm god. Later known by the Aztecs as the Pyramid to the Sun and the Pyramid to the Moon, the pyramid to the sun was situated directly over one of many caves. The cave was considered the place of origin, "where humanity emerged into being." In Malta, the temples from the fourth and third millennia B.C. were egg-shaped and were constructed underground with alters shaped like pubic triangles. Spiral motifs, snakes, lizards, and phallic forms were intermixed with plant life indicating the transformative, regenerative, and sexual celebrations of life. The inner chambers were often painted with a red ochre wash indicating the sanguineous womb. Many of these temples were constructed on top of or near sacred caves with sacred springs in their gardens. Often situated near the sea, the flowing waters were likened to the waters of the womb. At the temple of Tulum in the Yucatan, perched above the sea, murals depicting childbirth were painted on the walls so that women on pilgrimages of initiation could be instructed into the mysteries of birth.

Temples such as these were created in reverence of the earth. Their consecration was dictated by the power emanating from the inner sanctums of the earth's wombic chambers—the caves, the wells, the holy springs. Something must have been felt intuitively in one's core as prayer was offered to the womb, to fertility, and the promise of renewal. These places of worship concretized the sense of awe at the miracles of creation. It was in such places that the sanctity of the womb was honored—in the body of the earth, the body of woman, and all of creation.

Inside our bodies, we carry an imprint of the earth's life force and the electricity of the starry heavens. Spirit comes into matter only though the vehicle of our humanness. The cells of the body recall their journey through evolution, through ancestral genes, and through the waters of the birth canal. This miracle has so often been devalued in our modern world paradigm, that is, the revelation of creating life and how integral this is to the whole template of existence. It is no less than the key to our sense of relatedness to all of existence.

The impetus of the biological imperative is like no other energetic force except, perhaps, the certainty of death's beckoning. Birth and death co-exist at opposite ends of the powerful spiral of existence, both keepers of the labyrinth, different in form, but twins in the force of their intention. This is why many ancient tombs were also symbolic wombs or temples where the mysterious acts of birth, death, and rebirth were enacted—from which all life formed and all life ultimately departed. The earth's cycles of dormancy and fertility also reflect a woman's cycle—where the lining of the uterus is shed from the body during menstruation and gradually becomes receptive again to a fertilizing sperm during ovulation, when pregnancy is a possibility.

Honoring the womb of the earth, in whose belly reside the secrets of our origins, is encompassed in our biological or cellular memories. The prevailing wave of disassociation from source, however, or from the *omphalos*, is the tragedy we face both environmentally and spiritually. A dramatic disruption occurred sometime in our ancient past where the power of the womb, once revered, became something to fear and control by the doctrines of many of the new religions. Attention was shifted from the holy earth below to the heavenly father above.

Our modern world does not respect and protect the profound, life-affirming power of the sacred womb, and further the regenerative organism of nature. Instead we have embraced an efficient, scientific model, where one "part" has very little to do with any other part. Viewing all of life as a whole interrelated community isn't cited in our cultural encyclopedias.

The truth is, the brain cells of creation are contained in every living thing—they are small seeds holding a piece of the universal puzzle, awaiting a sprinkling of water and the sunlight beaming from cultivating eyes. The bloodlines to our ancestral origins forms a complex web of inter-relatedness, even though modern humanity has elected itself to the top rung of the evolutionary ladder.

Awareness and understanding take practice and diligence. We are offered the potential for communion and for sacred reverence of life but it is in the material of unmolded clay, awaiting the sculptor. This raw material requires attention, nurturing, and refinement. Why and how has it come to pass that we've abandoned this potential for unity in ourselves, amongst each other, and with nature? At its inception, all life is innocent and the interconnectedness is obvious. There is no question as a baby floats in its mother's womb, that if the umbilicus connecting it to its source of life is severed, the baby will die. The source, the nourishment and the life are all one and the same. Our environmental dis-ease has now become the barometer for humanity's lack of connectedness and life-sustaining support. The challenges we face in our time reflect a species out of balance with its larger womb—the earth.

The gift of consciousness is the birthright of our species. With it we form and design civilizations, create poetry and art, give meaning and explanation to our world and our lives. From it comes a will which can develop and destroy life. We've also been encoded with the powers of discrimination and a propensity towards developing ethics. Somehow, we've gotten communion confused with a selfish protectionism. Most species are fundamentally selfish; it's in their/our interest to survive at all costs. This self interest has been built into the mix. Selfishness now, however, becomes the modus operandi in the form of self-preservation to the detriment of nature because the relationship to source has been betrayed. A certain balance has been shaken loose from its base equilibrium. A stronger attachment to materialism has become a convoluted connection to some kind of sustaining source. Are we as a species forsaking the mother for more selfish ends? Clearly, there is now an urgency that we "grow up" and take responsibility for our home, and re-sanctify this place that is *still* the sacred womb of creation.

The imbalance in which we now live speaks of a sickness of spirit, a malaise of the soul—a disconnection from our source of nourishment. Many of the laws of the source have been forgotten; how and when is only conjecture. In the temple of Delphi, the oracular priestesses (named Pythias, after the sacred python, a symbol of transformation and regeneration) would sit on tripods over the chasms of the source. These high priestesses (inspired by the vapors rising from the crevasses in the earth, exuding from the Python) guarded the fertility and deep mysteries of the earth. The vapors or spirit induced the priestesses into a trance state, where their prophecies were spoken. The last prophetic words recorded by the Pythias, possibly indicating a "shift" was occurring away from a revering earth wisdom, were these . . .

> The Sacred Spring is running dry
> the carven wall has fallen in decay
> there will be no more prophesizing bay

The reverence for the holy cave of the earth, in the form of a temple, had diminished and these last prophetic words had begun to hollow into a voice deserted by spirit, *or* by humanity, which came first?

One of the primary functions of organized religion throughout human history was to offer explanation to the life-giving power of the womb and at the other end of the pendulum swing—death. This awesome mystery of the cosmic womb as life-giver and life-taker was somehow obfuscated by the dogma of the church. Later teachings hold that a heavenly god was the primary life-giving principle, not the sacred domain of the womb. Sadly, in our current environmental dilemma—the fecund, pregnant earth principle is likewise replaced by another kind of dogma—technology and the "efficient," mechanistic mode of being, with little regard for the sacred nuances of time and the body of matter.

Many indigenous people's cosmologies are based on a spiritual relationship to the body of the earth. Their spiritual practices often originated from a direct relationship with the natural world. A recognition of the sacredness of the womb and the spiritual protection needed during pregnancy was implicit. Indigenous spiritual wisdom addresses communion with the earth and protection of the seeds of life without separating humanity from the whole of nature. It is understood in Native American cosmology that failing to honor the laws of nature, dire consequences will result.

Isn't it time to renew our reverence for our own physical bodies as kin to the earth's body and human consciousness as a sacred well for life's mysteries? The sanctity of the womb's matrix is a metaphor for protection and caring for all of life. We must affiliate ourselves once again to that creative, renewing source of life emanating from the belly of matter—where all is inter-related and all has purpose. We must return with reverence to the holy temples of the womb, to the caves, the sacred chambers of Laussel, to the kivas—to remember the source of creation, the source of our privilege to *be*.

> The divine child . . . the divine principle of the universe at the
> moment of its first manifestation . . . was hatched out of an
> egg which came into being in the waters of the beginning—

hatched, . . . out of the void. (S)he reclines on the back of sea monsters, floats on the cup of water flowers. (S)he is the primordial child in the primordial solitude of the primordial element: the primordial child that is the unfolding of the primordial egg, just as the whole world is his(her) unfolding.[1]

Dedicated to my mother, Marcelle Weyl, and father, Julian Cowan.

—24—
Belonging to the Universe

David Steindl-Rast

(*From a lecture at the Earth Trust Foundation*)

Belonging to the universe. . . . I use the key word, *belonging*. . . . I suspect that most of us find the conflict of belonging somewhat problematic or else we wouldn't be here. And the problematical belonging begins long before it's a matter of belonging to the universe. It's a belonging on every level, belonging to our families, belonging to our communities, belonging to our country. On all different levels, belonging is for us today a key issue. And a problematic issue. Because our main problem today is alienation. We all experience alienation . . . in different ways. Beginning with alienation from ourselves, alienation from others, alienation from our feelings. The word *alienation* is almost the one word that sums up our problems, and between these two poles of belonging and alienation is the realm of life.

The realm of spiritual life. Because for me, spiritual life is not a separate department from life. But when I use the word *spiritual* it means, as the root meaning of the word it is, aliveness. Spirit, *spiritus,* is the life breath and so spirituality is aliveness. An aliveness on all levels in all areas, and that aliveness, that spirituality, moves between belonging and alienation. Or, you could say it is a move to ever greater aliveness and that means a move to ever deeper belonging. Or, an increasing discovery of that belonging, which we have from the very beginning.

As T. S. Eliot says in *The Four Quartets,* "We shall not cease from exploration. And the end of all our exploring will be to arrive where we started and know the place for the first time." And in a more explicitly religious context, Augustine says of this starting point, "God is closer to us than we are to ourselves in our heart of hearts." And yet he also says, "restless is our heart until it rests in You, oh God." And so from that starting point, which is home, we are already closer to the ultimate than we are to ourselves, to discovering this belonging.

Between those two points lies this longing to belong which continuously drives us on until we enter through what Eliot calls in the same passage, "the unknown remembered gate."

It us unknown and yet it is somehow remembered because it is where we start from. And that is our spiritual quest. And now to bring this home to ourselves, each one of us. And to explore it together and to see some of the implications of this quest to belong. Of this longing to belong to the universe and to that which goes beyond the universe to the horizon beyond all horizons. In order to do that together we have to appeal to our personal experience in which we have the deepest sense of belonging. And those are the moments which psychology calls peak experiences or peak moments. Everyone has these experiences and most people very early on in life, in childhood, quite frequently.

Even if it's only a split moment. Our whole heart goes out and says, "This is it." In the sense of "That's what I've always been waiting for." "That's what I've always been longing for." And that longing is a longing to belong. And for a split moment we belong. And then we lose it again. It's gone. And usually what triggers these experiences isn't even as spectacular as a sunset or a waterfall, it may just be the dimples in your baby's cheeks as you're looking at your baby. Or it may be the way a squirrel takes a nut from your hand when you are feeding it in the park.

Abraham Maslow, the great psychiatrist, in the middle of this century coined the term *peak experience* based on what is known as a mystic experience. The only reason why he switched is that in psychologic literature mystic experience sounded a little too mystic and it didn't quite sit. So he had to say something that was a little more kosher in the context of psychological literature.

All those books of mysticism are about the mystic experience. But, ultimately, this is inexpressible. So the one way in which it is somewhat accessible to us is through poetry. Kenneth Rexroth's "The Heart of Hercules" is obviously inspired by one of these experiences.

> Lying under the stars in the summer night
> Late while the autumn constellations climb the sky
> as the cluster of Hercules falls down the west
> I put the telescope by . . .
>
> my body is asleep only my eyes and brain are awake
> the stars stand around me like gold eyes
> I can no longer tell where I begin and leave off
> the faint breeze in the dark pines and the
> invisible grass
> the tipping earth
> the swarming stars have an eye that sees itself.

All this has an eye that sees itself. And of course, there is even a play on words because they have an eye and who is that I. I am that eye of the universe that sees itself.

All the different key words are translated into experience here, translated into image. The aliveness, obviously, the meaning. This is a meaningful moment. The belonging, "I can no longer tell where I leave off and where I begin." I merge. I merge with the universe. And it is in that sense truly a peak experience. Peak of awareness, peak of aliveness, and truly a mystic experience. It is not only a contact with nature, not only a contact with all that is, but a contact with the horizon beyond all that is. And then identification with that. It's a mystic experience by the most simple definition of the word mystic; an experience that must be personal; an experience of communion in the words of belonging. Of communion with reality. Yes, but with ultimate reality.

And every word in that definition counts.

Let's first look at it as this mystic experience, this encounter with spirit. Mystic experience in the sense of being ultimate; relatedness to the source of all there is. Relatedness to God, if you want to use this term God. I'm always hesitant to use God because you lay yourself open to so many misunderstandings.

What God means, if this word means anything to us, is that limitless belonging that we experience in our peak moments. All theistic religious traditions will agree that this is definitely the denominator for what we mean by God. That to which we ultimately, limitlessly belong. The reference point of this belonging. And in this experience it is no more than a direction. The direction of our belonging. God is the direction of our ultimate belonging.

Then, of course, the different religious traditions fill this in and there's a lot of exploration that you can do in that direction. But at least we have a common denominator and whether we want to use that term God or not, we know how it is correctly used.

But we don't need to use it. We can just speak, for instance, of all things and their course. The source from which all things come forth. That's what we are concerned with in religion. The ultimate source of all there is. Not things yet, but the source. There are things and then there is the source. Just as there is a little stream and there is the source of that stream. Well, the source is not the stream. Where it flows, it's already the stream. What is the source? The source is before it starts flowing. It's really . . . when we speak about all things and the source of all things, then the only "thing" that the source can be is nothing. If it's a thing, it's already something. It's not the source. If it's not a thing, but the source of all things, it's no thing. But not an empty nothing. Not just a denial of thing, but the source, that nothing out of which the fullness of everything comes. But I'm not concerned with the words, I'm concerned with personal experience; not things, but nothing. Because there are two things that matter in our lives: "things" and "meaning." And meaning is no thing. Meaning is nothing. And meaning is that source out of which everything comes. Meaning is nothing. You're not adding something to your house to make it your home. That's the meaning your home has for you. It's not something that you add to your house like you add a new room, or a new attic, or a new roof. You're adding nothing, but that nothing is what matters to you. So the source is that nothing and that is what religion is about. That source, that horizon, that from which everything gushes forth.

And in our peak moments, in our mystic moments, because we are all mystics, we experience this *meaning*. Meaning beyond all things. Now there is a question that is very important to many people, and that is the question of our relatedness to the ultimate source. And how do we get from that mystic experience to the religions? How do we get from religion with a capital *R* to the religions?

Inevitably our intellect does something with this experience, reflects on it, tries to understand and interpret it. And the moment the intellect interprets what's happening we have myth. And to the extent to which we can understand that poem by Rexroth as a religious poem, and it certainly is, to the extent to which it is the expression of a religious experience, it ends up in a myth. And it's a very beautiful myth that deeply speaks to us, and it is the myth of the universe looking at itself through my heart, through my eyes, with antecedents in more ancient myth; the poetic expression of an inside that is too deep for any other expression. There are insights into the human condition that are so deep that no other language will be strong enough to carry it. And we hit those experiences when we are in love, for instance. And that's why we start writing poetry when we are in love or at least expressing poetically or waxing poetically. Only poetic language can carry it—the poetic expression for an inside that is too great for any other kind of expression like logic, or abstract thought, so that's the first thing that we inevitably get, myth. And myth is an element of every religious tradition.

Secondly, our will does something with this experience. Willingness. Not our willfulness. Our willingness. Our willingness is always after what's good. We can only will what is really good for us. And so our willingness says, our heart says, "This is terrific. This belonging, in the sense of belonging. This is how I would like to live." And immediately you get ethics, which is the second element of every religion. What all of the ethical systems in the world have in common is that they say, "This is how one lives when one belongs together. That is how one behaves towards animals, toward plants, towards other human beings, towards the whole universe, when one is at home in this earth household." So that's the second element.

And the third element is that our feelings, not only our intellect, our will, but our feelings come in and our feelings celebrate this experience. And this is terrific. And the moment the feelings celebrate the mystic experience, the sense of belonging, we have ritual. And every ritual is a celebration of belonging. Anything from a birthday party to a graduation, to a religious ceremony, to a peace march, everything, every ritual, whether you create it, or whether it's ancient, is always a celebration of belonging. And here we have the three elements that belong to every religion with a small *r*: myth, ethics, and ritual. Of course, they can go wrong; myth can turn into dogmatism. Ethics can turn into moralism, and ritual can turn into ritualism. But when it goes well, then myth will again and again lead us back to the mystic experience. Ethics will be an overflow, will be the social realization of inner achievement and fulfillment, lived out in community. And ritual will be the celebration of it. The joy of life. All of life can become that ritual.

Let us look at this once more. This experience, but not primarily under the religious aspect, not primarily under the aspect of spirit, but now under the aspect of nature and of our dealing with nature. And here you have first what you could call cognition. What nature is, is that which gushes forth without our doing anything. That's what nature is. We spoke before of the source and then we speak about that which gushes forth. And that is nature. And the very term, *nature*, the very root that stands behind this means this gushing forth. You have it in nativity where something comes forth. Our first contact with nature, in this poem, for instance, or in any such situation when we lie under the starry skies is cognition. That is still wordless. And notice that very word, *cognition*, has also the same root in it. It's *co-* that means together and then again this gushing forth. So what we experience in cognition is that we are part of that which gushes forth. All this as Rexroth says here, all this gushing forth of nature has an eye that sees itself. And that's cognition. We are that eye. Cognition is a recognizing, redoing of that first moment. So here it is done in poetry. That recognition. And you have it here in this poem, for instance. When the poet says, "my eyes and brain are awake . . . the stars stand around me like gold eyes . . . I can no longer tell where I begin and leave off . . . the faint breeze in the dark pines and the invisible grass . . . the tipping earth . . . the swarming stars." All this is recognition. It is expressed poetic language. That recognition of the original cognition.

The third possible step that we can combine with this encounter with nature is reflection. The first was cognition, then recognition, and now reflection in this recognition, and that is science. But there are two very different possibilities for science. We nowadays think only of one and that is exact science, or Cartesian science. But there is originally another kind of science and you may call this chemical science. Alchemy is a reflection on the poetry of the encounter with nature. It includes the personal and the religious.

And then you have the exact science, purged of everything that is poetic, everything that is personal, and everything that is religious. It must be purged. In other words, the science that we call the exact science is not a reflection on the recognition; is not a reflection on the poetry, but goes directly back to the original encounter with nature and tries to do all this unpoetically. And that is why our science, where all the personal is left out, suddenly becomes totally impersonal and we feel totally alienated.

When science is the only recognized access to nature or to reality—and we have a tendency to view it that way—but there is another possibility: we can put science into a much greater context. And every scientist who really reaches a certain stature does that anyway. As a human being realizes that science is not all of life. It is part of life. It is a very important part of life. It can give us important access to life, but it is not enough. Life is greater than science. And that greater aspect is the poetic aspect, is the religious aspect, and it needs to be added.

There is a parallel to this and now I'll go back for a moment to what I said about the intellectual understanding of the mystic experience. I said it leads to myth. The intellect tries to interpret this encounter with ultimate reality, this sense of belonging, and arrives at myth. There are other people who would say it leads to theology. Does it lead to theology? No. Theology stands in the same relationship to myth as science stands to the poetic encounter, to the poetic expression of our encounter with nature. And so theology is really something like literary criticism of our poetic myth. It's as valid as literary criticism is and it also runs all of the risks that literary criticism runs. We all know that literary critics start out with poetry and can do something very valuable there, but they have the tendency to get so caught up with literary criticism that after a while they no longer really talk about poetry. They talk about literary criticism to literary critics. In very much the same way theologians start out with talking about myth and ultimately about mystic experience and end up talking with other theologians about theology. And that leaves most of us left out in the cold. But we are, if all goes well, concerned with what really matters, and that is the mystic experience and not the reflection on it and the reflection on the reflection.

So what we need to do to bring back that sense of belonging, to find ourselves back there where we really want to be, is put what we have, our theology, if you will, our science, if you will, into this larger context of life—the mystic context. There is one other aspect that I would like to explore briefly with you, because it also belongs to this. I emphasize meaning so much. That these peak moments, these mystic moments are also moments of meaning. It might be helpful for our discussion to clarify what we mean by meaning and purpose. Because in our everyday language we often confuse these two and that is a very difficult and dangerous confusion. When something becomes meaningful to us, we must give ourselves to it. We must allow it to take hold of us. When it is a matter of purpose, we must take things in hand and control them. So it is a totally different inner attitude that we have towards purpose and towards meaning. You know this from your own experience. When you want to achieve a particular purpose you must control things. When does meaning happen to you? When you give yourself to it. When it does something to you. And that meaning whenever it happens has three aspects, and one is, there is something that has meaning. It may be a situation, it may be a thing. It may be an encounter. It may be a landscape. Whatever it is it is something. And we say of it . . . it speaks to me. It tells me something. That's why it has meaning to me. And so the first thing that is always there whenever we talk about meaning is something that speaks to us. But there is something else. Just as we spoke of all that is gushing forth there is that from which it gushes and that is silence. And there cannot be meaning without silence. Because there cannot be word without silence.

We all know very well the difference between chitchat and real dialogue. And real dialogue is not an exchange of words but is really an exchange of silences, often by means of words and sometimes without words. So it's the silence that is another element of meaning. Without silence, nothing can become meaningful. And then there has to be a third element, and that is understanding. Because you cannot imagine that there is a world and there is silence if there is no understanding. Well, if there is no understanding again there is no meaning. These three belong together and they're a very mysterious threesome because you have no overarching concept of what these three are together. If you have apples and oranges and bananas, you have three kinds of fruit. But if you have silence, word, and understanding, you have three what? You have no overarching concept here, you have three worlds. And yet they're extremely important for us in every moment in which we find meaning. It is another aspect of this longing to belong. The religious quest of humankind throughout the ages is this quest for meaning, an exploration following the word, the one most of us are familiar with because it is the "word of God." This ultimate source that is no thing gushes forth and we can understand that. It speaks to us. In the beginning was the word. That's what the Western traditions have in common. But there is a whole other world of religion, one of which is Buddhism. And there silence is as central as the word is to us.

To take one example; the sermon of Buddha which would correspond to what we call the Sermon on the Mount in the Christian tradition—which is lots of words, three chapters of words in Matthew—is a wordless sermon. And all that Buddha does is he holds up a flower. How can anybody prove that they have understood? If they say anything it is obvious that they have missed it. Because it is about silence. So how should anybody prove that they have understood; well, one smiles. And Buddhists say at that moment the tradition of Buddhism was passed on from the Buddha to his successor, who was the one who smiled. And ever since, the tradition of Buddhism with all its books and sutras and all the rest has been passed on in silence because it's the silence that counts. And I remember my Buddhist teachers whenever we got into any discussion—and they have lots to say about Buddhism just like the mystics have lots to say about mysticism—but whenever my Buddhist teacher would get carried away, he would somewhere in the course of instruction catch himself and say, "Well, I've been talking again. I've become a Christian." And whenever I came to a point where I would have to say, "Well, you know, I can't explain it any further. It's what you call a mystery." The very word *mystery* comes from *wu a yen* which literally means to shut up. So you can't say anything. It's a mystery. So, then, at that moment when it was very tense and I was listening and trying to catch it all. When I said "It's a mystery," he already knew what was coming and he would relax and he would smile and say, "I understand. I understand." We had reached the Buddhist realm of the world religions. And then there is this third realm of understanding, and that is this understanding. It is the process by which you give yourself so deeply to the word that the word takes you where it comes from. The word takes you into the silence. All that gushes forth takes you into its source. The silence comes to our word. That process, that dance, that movement where it goes from the silence into the word and from the word leads again into the silence, that is understanding. And that is as central to Hinduism as the word is to the Western traditions and the silence is to Buddhism. And in Hinduism, the key that would unlock for you the understanding of any Hindu word or teaching is, "Atman is Brahman and Brahman is Atman." That means that manifest, divine reality is the unmanifest, and the unmanifest is the manifest. Or the word is the silence and the silence is the word. Or that which gushes forth before it gushes forth. The two are one.

And to understand that word is silence and silence is word that is understanding. But it's not an understanding somewhere in the head but an entering into this process. So we have actually that quest for meaning focusing in three different main religious traditions, on three aspects of meaning. Word, silence, and understanding. And the three are one because you cannot have the one without the other. In the Western tradition you certainly have the silence and the understanding because you cannot have word without them, but they are marginal. And in Buddhism you certainly have word and understanding, but they are marginal. And in Hinduism the word and silence are marginal to the understanding. In the Christian tradition these three elements were of course also recognized as divine. And so you speak of the *logos* which is the word. And you speak of God as the abyss of silence, as C. S. Lewis says, "God is that abyss of silence into which creatures can drop down their thoughts forever and ever and never would they hear an echo coming back."

And you speak of a spirit of understanding. God's own self understanding within us. The God who is closer to us than we are to ourselves. We cannot understand God except through God's self-understanding. And so all three are there. And this is what Christians call the triune God and they speak very early on of the round dance of the trinity. The word comes out of the silence and through understanding it turns into the silence. And today, we see this round dance of the trinity because our perspective has become worldwide. We see it reflected, if you want, in the round dance of all the great world religions with one another. And we are all engaged in the same quest for meaning.

But as long as you stand on the outside, and that is one of my favorite images, as you stand on the outside of this round dance, it will always appear to you that those closest to you are going in one direction and those furthest away are going exactly in the opposite direction and there is no way of verifying this vision until you get in.

And the moment you hold hands and you move with it it is absolutely obvious that we are all going in the same direction. And it doesn't make any difference where you get in. It doesn't make any difference. Get in wherever you are because that's the only place where you can get in. You can't very well get in at any place where you are not. So wherever you are you get into this dance, and before you know it you see we are all dancing with one another.

—25—
Creation Spirituality

Matthew Fox

(*Extracts from the Schumacher Lectures 1991*)

What would an ecological religion look like? Humankind has been involved in a gross desacralization of this planet, of the universe, and of our own souls for the last three hundred years. Here lies the origin of our ecological violence. Can we recover the sense of the sacred?

Religion's future is not in religion itself. Religion has to learn to let go of religion. Meister Eckhart said in the fourteenth century, "I pray God to rid me of God." In order to rediscover spirituality, which is at the heart of any authentic and healthy religion, we have to be free of religion. This is a paradox. Spirituality is the praxis of the heart, the praxis of our living in this world. It means dealing with our inner selves and not just living on the level of our outer organizations.

E. F. Schumacher in his prophetic way, named this issue in his epilogue to *Small Is Beautiful* when he said, "Everywhere people ask, 'What can I actually do?' The answer is as simple as it is disconcerting. We can, each of us, work to put our own inner house in order. The guidance we need for this work cannot be found in science or technology, the value of which utterly depends on the ends they serve. But it can still be found in the traditional wisdom of humankind."

Thomas Aquinas in the thirteenth century said that "revelation comes in two volumes—the Bible and nature." But theology, since the sixteenth century, has put so much emphasis on the word of the Bible, or the word of the professor or the word from the Vatican—we have put all our eggs in the basket of the word—the human word. We have forgotten the second source of revelation, that of nature itself.

Meister Eckhart said, "Every creature is a word of God and a book about God." In other words, every creature is a Bible. How do you approach that biblical wisdom, the holy wisdom of creatures? You approach it with silence. You need a silent heart to listen to the wisdom of the wind and the wisdom of the trees and the wisdom of the waters and the soil. We have lost this sense of silence in our obsessively verbal culture. Schumacher said: "We are now far too clever to survive without wisdom."

Lester Brown has put in one sentence what a lot of Green activists and scientists are realizing today: we only have twenty years left on this planet to change our ways. We, therefore, need to explore our inner houses as urgently as possible. When we use the term "inner house," remember that the soul is not the body. The inner house is not this little thing situated in the pineal gland which Descartes called the soul. All our great cosmological mystics—Hildegard of Bingen, Thomas Aquinas, and Meister Eckhart—have said that the soul is not in the body but the body is in the soul. The body is an instrument of our passions, of what we really care about. Of our grief, of our wonder. Exploring the inner house of our soul means listening to the deep self. This exploring of the inner house is not just one's personal inner house, but the inner house of our communities, the inner house of our nations, the inner house of our gender, the inner house of our species. In other words, the inner house is not just part of an individual; our way of life contains an inner house. It is because we are violent inside that our environment is dying all around us. The nest in which we live, we are fouling.

So, this exploration of our inner house has everything to do with the ecological era. The word *ecology* means the study of our home. Ecology is not something "out there." We are in nature and nature is in us. We are in the sacred home and the sacred home is in us. The sacred wilderness is not something just out there, only to be found in our national parks. There is a sacred wilderness inside every one of us and it needs our attention. We are out of touch with the sacred wilderness of our passions; that is why we see such devastation all around us.

Religion in our time must undergo the recovery of its mystical tradition. The British monk Bede Griffiths has recently said, "If Christianity cannot recover its mystical tradition, then it should simply fold up and go out of business. It has nothing to offer." To that I say Amen. The churches are empty and the souls of our young people are being possessed by despair. It does not have to be this way because we can recover our mystical tradition, for there is a mystic in every one of us, yearning to play again in the universe. When you develop mysticism, then you are in the field of developing prophets.

As the American philosopher William Hocking said, "The prophet is the mystic in action." Carl Jung said, "Belief is no substitute for experience." The mystic in every one of us trusts his or her experience of the divine in nature, which opens our hearts. And when our hearts open, the divine comes through. This trust of our experience is the basis of all mysticism.

The psalmist says, "Taste and see that God is good." Mysticism is about tasting—there is no such thing as vicarious mysticism—The Pope can't do it for you, your parish record cannot do it for you, you can't rent a mystic, not even in California! We are moving into an era where we must all take responsibility for our mystical lives, calling on the wisdom of our ancestors and the wisdom of our communities, including, of course, the wisdom of the non-human community which is constantly feeding us its revelation and truth, and teaching us that we can still taste and see that divinity is good.

Aquinas put it this way, "The experience of God must not be restricted to the old or to the few." We do not need professional mystics, we all need to wake up to our own mysticism.

Gregory Bateson in his book *Steps to the Ecology of Mind* says, "The hardest thing in the Gospel is that of Saint Paul addressing the Galatians when he says, 'God is not mocked.' This saying applies to the relationship between humanity and ecology. The processes of ecology are not mocked." In other words, the earth has been keeping a ledger about the ozone layer, the pollution of the atmosphere, and the deforestation, because the earth will not be mocked. As Hildegard of Bingen said, "There is a web of justice between humanity and all other creatures." She says that if humanity breaks this web of justice then God permits creation to punish humanity. Creation is already responding with cancer and leukemia. The earth is not mocked.

Bateson then analyses the three main threats to human survival. The first is technological progress, the second is population increase, and the third is errors in the values and attitudes of Western culture.

If we have only twenty years left then we have to start awakening the masses and the route for awakening the masses is the sacred religious traditions. They would affect our conventional attitudes toward our bodies, toward health, toward wholeness, toward the sacredness of all creatures in whole new ways. I am talking about a religion as if creation mattered. It will teach us spirituality, it will teach us ways to live non-violently with ourselves and therefore with others.

To retrieve this kind of religious wisdom, one thing we need to do is to look at our own spiritual traditions which have often been condemned. The greatest Saint of the creation tradition, Hildegard of Bingen, was ignored for seven hundred years. Francis of Assisi was sentimentalized and put in a birdbath! Thomas Aquinas was condemned three times and then made a Saint. Meister Eckhart was condemned and is still on the condemnation list six hundred years later. Julian of Norwich was ignored. Her book was not published until three hundred years after her death. The creation-centered Celtic people in the seventh century had their nature-mysticism smothered at the Council of Whitby.

These people were steeped in a creation-centered spirituality. A spirituality that begins with original blessing instead of original sin. The idea of original sin is radically anthropocentric. Sin is only as old as the human race. I deny the prominence that the Western church has given to original sin. It has fed this anthropocentrism, it is so egoistic to think that religious experience begins with our sins.

I believe that religious experience begins with awe and wonder—that is the first step in the spiritual journey. Awe is the beginning of wisdom. There can be no compromise on this truth. The first step toward spiritual revolution is to recover awe and wonder in our time, and this is in fact a rather easy task, because we are being given a new creation theory from science itself; a new cosmology about how our species got here, how this planet got here. No one can hear this story without being filled with awe and wonder.

To hear that all the elements of our bodies were birthed by a super-nova explosion five and a half billion years ago, which unites us with all the elements in the universe, is awesome. To hear that in the first seconds of the fire-ball, decisions were made eighteen billion years ago on our behalf, that the temperature of the fire-ball had to be within one degree of what it was for this planet to evolve, is awesome. And that is how it happened. When you hear these stories tumbling out of the mouths of scientists, it is no wonder that they are leading the way to mysticism today. I hear the echo of Julian of Norwich in the fifteenth century who said "We have been loved from before the beginning."

Unconditional love is the first lesson of the new creation story, and it is the lesson of all the mystics as well. When you put science and mysticism together, you have a new cosmology bubbling up. When the artist gets on board and tells these stories in song and dance, music and ritual, then you have a renaissance; a spiritual rebirth based on a new vision. All this awe is the starting point for a spiritual life. Aquinas put it this way, "All things have been made in order that they imitate the divine beauty in whatever possible way. Divine beauty is the cause of all states of rest and motion, whether of minds or spirits or bodies."

We have not heard the word "beauty" held up as a theological category for three hundred years. Descartes, the father of Western academia and of Western science, built a whole philosophy without any mention of beauty or aesthetics. The last time we had a cosmological spirituality in the West was in the Middle Ages when there was a celebration of the beauty of God. Francis of Assisi said "God is Beauty," and Aquinas taught that all of us share and participate in the divine beauty. The ecology is speaking to us at the most radical level, because we are all falling in love with the earth. The first step in the ecological journey is to fall in love with the beauty of this planet, so that we will defend it and liberate it when injustice threatens it and abuses it.

Rabbi Heschel said, "Just to be is a blessing, just to live is holy." Heschel explains that there are three ways in which we humans respond to creation. The first way is to enjoy it, the second is to exploit it, and the third way is to accept it with awe. Our civilization in the West has never practiced the third way—at least not as a civilization in the last few centuries. Awe was taken out of the classroom, because Descartes defined truths as ideas, so our entire educational system is modeled on ideas.

Once a scientist said to me, "For the last twenty-one years I have been doing nothing except locking myself in a laboratory at Stanford University examining the right hemisphere of the brain. I am now ready to publish my findings, which are that the right hemisphere of the brain is all about awe." We have become a one-sided civilization because we have not been nurturing the right side of our brain, which is the sense of awe. Our species has to redefine our relationship to nature, including the wonder of our being here with the sense of awe.

This is why I dismiss the stewardship model in theology as being totally inadequate for the ecological era and an environmental resurrection. The stewardship model tells us that God is "out there." God is an absentee landlord and we are here to do God's work. Therefore, we have a duty-oriented morality, but you cannot inspire people by the concept of duty. You only make them feel guilty, which tires people out. The idea of duty-morality goes back to Kant, it is part of the Enlightenment. Let it go.

Aquinas said that you change people by delight. The proper model for theology for an ecological era is not stewardship which reinforces duality; the proper model is mysticism, the Cosmic Christ and the Garden in the Song of Songs, where we realize that God *is* the Garden. That God is expressed in each "word" that the plants are, the trees are, the animals are. And when these are being jeopardized, God is being crucified. When they are splendid and healthy, divinity itself is radiating its *doxa*, its glory. The Cosmic Christ is radiating its glory in the glory of nature.

A shift from the duty-oriented stewardship ethic to mysticism and the Cosmic Christ is the basis for an ecological spirituality. This is the home, the "*ecos*" in which we live, it is the divine home. "God is here, it is we who have gone out for a walk" (Eckhart). Divinity is everywhere, but our eyes have to learn to see again.

Another dimension to ecological spirituality is the very word "environment." It comes from the French word *environ* which means around. The proper theology for our era is not about a God out there somewhere, it is about a God who is around us, as Julian of Norwich said, "Who is completely enveloping us." It is a very maternal image of God and as Hildegard of Bingen said, "You are hugged by the arms of the mystery of God."

This is pantheism; it teaches us that everything is in God and God is in everything. That is the proper way to name our relationship with the divine. Mechtild of Magdeburg, a social activist and feminist of the thirteenth century, said that "The day of my spiritual awakening was the day I saw, and knew I saw all things in God and God in all things."

That is the day we grow up spiritually, and if we don't have the resources in our culture to assist us in that growing up, if we do not hear the teachings of the mystics, then we have to go out and demand it. Just as we need to take back our bodies from the medical industry, so too do we have to take back our souls from the professional priests who are not doing their job to the extent that they themselves have been wounded in seminaries and reductionist education.

Lester Brown uses an important word when he talks about "inertia." The medievals had a word for this which is very deep. They called it *acedia*, which means refusal to begin new things. Acedia includes depression and sadness. Our spiritual tradition should address acedia. Because that is the issue; how do you awaken the masses? How do we awaken ourselves? Paul put it this way in his letter to Corinthians, "The sadness of the age is busy with death."

I was so aware of this when the American government aroused the people and sent four hundred thousand human beings half-way around the world with an untold tonnage of weapons. They aroused the people, but over what? Going to war. We have not managed to arouse our people over the despair in the cities, the treatment of the soil. It is just as Paul said, "The sadness of the age is busy with death." We only wake up to see death as our entertainment. Death is news. Death is now about the only thing that arouses us.

Aquinas said, "Despair comes from the loss of belief in our own goodness, and the loss of awareness of how our goodness relates to the divine goodness." In other words, it is the theology of blessing that is the proper response to acedia or inertia. It is when we get excited about the goodness of things that we are prepared to act for life and for the earth. It is like "falling in love." We have anthropocentrized "falling on love," we think it is something you do to find a mate for the rest of your life. But it is much more than that. We could fall in love with the galaxies. We could fall in love with species of wild flowers. We could fall in love with fishes and plants, trees, animals and birds, and with people. This capacity for being in love has no limit, and all of it is about experiencing blessing.

Aquinas said, "To bless is nothing else than to speak the good. We bless God and in another way God blesses us. We bless God by recognizing the divine goodness." So, we must take some time to meditate on the intrinsic goodness of things, the goodness of the forests, of healthy clean air and water. "God blesses us by causing goodness in us" (Aquinas). We have to take time to meditate on our own blessing, on how we are uniquely good. There has never been another collection of DNA like ours in the history of the universe and every single person is a unique expression of the Cosmic Christ. As the mystics say, everyone is a unique mirror of God. That is why there is such a diversity of creation to delight divinity. That is why all creatures are here—for the sake of delight.

We must let go of the Original Sin ideology, that has us growing up with shame and guilt about being here, which in turn creates compulsions of perfectionism and more guilt. The fact is, my friends, all creatures are imperfect— let us celebrate that. Divinity purposefully matched our imperfections with one another, so we need one another and that way we build relationships with one another. There is glory and beauty within the imperfection. Every tree is beautiful but if you go up close, it has its dead ends, its knots, and its broken branches. We are all that way too, and there is no shame in that. The shame is in wallowing in it and not paying attention to our goodness. Aquinas said that the sin of acedia, the sin of inertia, is a sin against the commandment to enjoy the Sabbath. The word Sabbath meant that God spent the Sabbath delighting in creation. We have to recover that sense of delighting in creation. When we delight in creation it is spiritual ecology; is it *Via Positiva*.

The second path of the spiritual ecology is *Via Negativa*, the way of darkness, the way of despair and grief. Our hearts are daily broken with facts about what humankind is doing to the earth, despair and grief hits us, but the first thing to do is to pay attention to that grief, and to let it be. To journey with the grief, to journey into the darkness. The mystics call it the dark night of the soul. Today our whole species is involved in the dark night of the soul, but that is not necessarily a bad thing; it can be the beginning of radical conversion, the beginning of new life.

Bede Griffiths says in his book *The River of Compassion*, "It is significant that the experience of despair is a yoga. Despair is often the first step on the path to spiritual life and many people do not awaken to the reality of God and the experience of transformation in their lives until they go through the experience of disillusion and despair."

As a civilization today, we are going through this disillusion and despair, and we need the mystical tradition to support us, because God is found not only in the light and glory of creation, but also in absolute darkness.

I have spoken about the path of creation and delight and also about the path of darkness. Now we turn to the third path which is that of creativity. A rebirth of creativity comes from delight and after the darkness. After the crucifixion comes the resurrection, the new birth, the surprise. Today we need to give birth to new virtues in many areas of our civilization.

Traditionally in the West we have political virtues, domestic virtues, and civic virtues. Today we need to have *ecological virtues.* For example, vegetarianism or semi-vegetarianism is an ecological virtue. There is no longer any excuse for a human being in the so-called first world not questioning his or her amount of consumption of meat. In fact, if North Americans alone were to cut back just ten per cent in their meat consumption, sixty million humans would eat today who were starving. The amount of land, water, and grain we are using to feed an addictive meat habit is simply unsustainable in our time. I am not saying that everyone *must* convert to full vegetarianism, but certainly we *can* cut back ten per cent and move on from there.

Another ecological virtue is bicycling, car-sharing, or walking to work. Recycling itself is an ecological virtue. Learning the sacredness of water and reverence for water is another ecological virtue. There are simple ways to learn reverence. Here is one I learnt from a Native American Indian, years ago; if you want to learn reverence for water, go without it for three days. After that, with the first sip you will rediscover the sacredness of water.

We must recreate our own entertainment at home and in our neighborhoods. The arts of conversation, gardening, drama, music, and tree-planting are delights in themselves. We have turned our entertainment to the television. For a long time I have maintained that if you are going to have a television in a house with children, for every hour that the kids watch television, they should be asked to put on their own show as well. We need to rediscover the arts of feasting together, and enjoying one another's company. Study is a spiritual praxis. To study the new creation story and then to put it to drama, ritual, and music is a spiritual ecology. To study the crisis of the forests, the abuse of animals, to study one's own history. Political organizing to defend creation, including civil disobedience when necessary, is ecological virtue. Another one is to make rituals: to celebrate sacred times and sacred places and the sacred beings with whom we share this planet. Rituals are how people have always passed on their value systems to the young. We need a revolution in ritual today. Ritual worship has become as boring as government or school! We need to bring our bodies back to ritual. Because you pray with your breath and your heart, not with books. Prayer is about strengthening the heart. We need people who can lead us through those prayers in new ways and traditional ways.

Art is the basic way to find the wisdom of our hearts. We could be putting our whole species to work today if we honored the artist as a spiritual director, which is what the primary role of art is, and always has been, in native traditions everywhere. I met an aborigine a couple of years ago. She said, "In our culture, everybody works four hours a day, and the rest of the day we make things." What is it they are making? Rituals, conviviality, beautiful costumes, music, and food for the feasts that follow the rituals. It is in ritual that the community heals itself, enlightens itself, brings forth gifts from everyone to celebrate and to let go.

The addiction to avarice and greed is deep within our civilization. It is built into the very structure of capitalism; this quest for more. Aquinas says, "The greed for gain knows no limit and tends to infinity." Avarice is not a problem of materialism, it is a soul issue, it is our quest for the infinite, but is has been misplaced. Consumerism cannot satisfy us, and this is why we are always looking for the new model next year; it is an infinite progression in the consumer addiction.

What is the answer to avarice? The answer is to put forward in our educational systems, in our religious traditions, in our political and economic arenas as absolute priorities those areas of the human quest for the infinite that are authentic and that can be satisfied in this life. We don't want to put down the quest for the infinite. Aquinas names three ways of the authentic infinite. The first is the human mind, "One human mind can know all things. It is capable of the universe," and his proof of this is that you never learn too much. So to feed the mind is to combat avarice. The second way in which we are infinite, says Aquinas, is in our hearts. "We can put no limit on the amount of love that one human heart is capable of." The third way in which we are infinite is through the human hands. "Connected to the human imagination, the hands can create an infinite variety of artifacts. Consider how, in the whole of the history of the human race, no two musicians have written the same song, no two painters have painted the same painting; an infinite capacity for creativity. If we want to remake our civilization, we must remake it around what is the spirit in us—mind, heart, and hands.

When authentic spirituality leads, religion will follow. If religion cannot make the paradigm shifts, if it cannot let go of religion itself and its dated sociological forms, then its value, in the West, is about on a par with the value of the Communist party in the Soviet Union. If it cannot relearn its own spiritual and mystical tradition, if it cannot touch our hearts and our bodies again, if it cannot teach a blessing consciousness in relation to today's new creation story from science, if it cannot teach us ways to journey into our shared grief and darkness, if it cannot teach the ecological virtues we need to survive, if it cannot offer us renewed forms of worship, if it cannot teach us about sins of the spirit like acedia, inertia, and avarice with as much enthusiasm as it has taught us about sins of the flesh, if it cannot apologize for its own sins toward native peoples, toward the earth, and toward women, if it cannot lead the way in bringing forth the wisdom of all the world religions, then, my friends, the young will grow old very quickly, and when that happens, a species dies. Given the responsibility of our species today, if that happens, we will bring down many other species with us. However, if we can rediscover a spirituality as if creation mattered, we will have a renaissance, a rebirth of civilization, a reinvention of our species based on a spiritual vision.

Part IV

—The Future of Nature—

—26—
The Fate of the Earth

Miriam Therese MacGillis

*(On May 28, 1986, Sister Miriam MacGillis, a Dominican sister from Caldwell,
New Jersey, spoke to an audience in Santa Rosa, California. Her presentation was sponsored
by the Department of Religious Education of the Diocese of Santa Rosa, Catholic
Community Services, Beyond-War, Sonoma Action for Nuclear Disarmament, and
St. Leo's and St. Eugene's Catholic parishes.)*

It is no accident that we've been born in these times, that we find our lives
unfolding now, with our particular histories and gifts, our brokenness, our
experience, and our wisdom. It is not an accident. In talking about the
fate of the earth, we know that its fate is really up for grabs. There are no
guarantees as to its future. It is a question of our own critical choices.
Perhaps what we need most is a transforming vision, a vision that's deep
enough, one that can take us from where we are to a new place; one that
opens the future up to hope. More than anything, we must become people of
hope. That's what I hope this reflection will be about.

I'm going to be speaking from a context created by Thomas Berry. He
is a Passionist priest and the president of the American Teilhard Association.
Teilhard de Chardin was a Jesuit priest who brought together in his own
psyche the insights that come out of the history of the Judeo-Christian tradition with the contemporary scientific understanding of evolution. He put
both together and integrated them into a world view that has probably done
more to shape the modern world than any of us yet understand.

Teilhard de Chardin died in the 1950s, before the ecological crisis. He
died before we had access to the image of the earth from space and before
the more available perception of the earth as a living system such as made
popular by James Lovelock. Thomas Berry has tried to take Teilhard's
work and bring it up into the present.

Often we question the fate of the earth and the critical state in which it
is now unfolding, and conclude that humanity is undergoing a kind of moral
failure. It's so easy to read the signs of the times and blame ourselves for
our lack of spirituality or religious fervor, or simply believe that we're just
an inherently greedy, selfish, destructive species. We tend to conclude that if

we could only come up with a religious revival, everything would be all right because our crisis is one of a moral and ethical nature. I'm not saying that there aren't levels of truth in that. But I think that there's really another area to consider before we make that final conclusion. That's what I'd like to suggest we do now.

Thomas Berry would interpret the crisis as a crisis of cosmology. Now what does cosmology mean? It simply means that all people have stories wherein they describe how the universe was made. All peoples on the planet have such a story; it's their origin story. Western culture has a genesis story. These stories reflect people's observations and conclusions about the origin and nature of the world. At this point in time, we find ourselves in a crisis that has to do with our original story, not just a crisis of evil. When we look at it in this way, we can begin to see that the future does open up to hope. It gives us the capacity to rethink our origin story with an expanded view of it. It gives us images that are positive. We're able to live out of those images. If our images of the future are negative because we conclude we are an inherently destructive species, or these are the end times . . . if we live out of those images, we're going to bring them about. Our images of the future are self-fulfilling. It's imperative to read the signs of the times in broader ways constantly, to deal with the signs of the times and allow the pain of them to come into us, and not be paralyzed by them. Above all, we can't deny them.

There's an extraordinary short poem by Wendell Berry that has touched this for me personally.

> In the dark of the moon,
> In the flying snow, in the dead of winter
> War spreading, families dying, the world in danger
> I walk the rocky hillside, sowing clover.

That kind of hope, that sense of significance of doing things for life, is what hope and meaning is about. It reminds me of a story of an old woman in the Mid-East who planted a date. When you plant a date, you know you're never going to eat the fruits of the tree, because it takes about eighty years for a date seed to grow into a tree with roots deep enough to take that scarce water and bring it up to the surface and produce the fruit. If you understand this process, you can make the commitment. You know that in the eighty-year period, date trees are buffeted by sandstorms and windstorms and all kinds of impacts on their growth. For the most part, the tree could look as if it is dying during those eighty years. If you did not understand this, if you didn't understand the *process*, you could easily make a judgment about the severity of its condition and cut it down. You have to live out of the image of what is going to happen, and that's what I'm talking about, living out of our images of hope.

A new perspective coming out of cosmology can enable us to do that. Hope is a choice. We make a choice to hope. And once we do that, it can make all the difference in the world about what we do. So let's look at this idea of cosmology and see how it connects to our present world crisis.

A people's cosmology, or origin story, predates everything else they create, their culture, religion, economics, politics, whatever. We can understand this better by way of comparison. For instance, the Native peoples who lived in this area before our European ancestors were operating out of a cosmology. Their experience of the world helped them shape an origin story. That story gave them a sense that the Great Spirit, the Divine Creator force, lived *inside* the universe; lived *in* the earth. Because they believed that, and because they believed that the earth was infused with the presence of the spirit, out of this understanding they evolved their religious systems, their manner of worshipping and coming in contact with the Divine. They shaped their morals, their ethical systems, and their economic systems out of the same beliefs. If the spirit was inside things, they concluded that all living beings were relatives of the spirit.

Native American cosmology reflected in a concrete, incarnate way the unique power of the spirit being transmitted through each creature. So, for instance, to take the feather of the eagle and to wear it was a way of being in touch with the divine power that the eagle reflected. They understood that the water and the trees were relatives, that the only reason they breathed was because the breath of the Spirit was in them. Thus, that affected the people's relationship to what was living.

This explains why it was impossible for them to invent an economic concept like private property. They lived here for seven to ten thousand years and it never dawned on them to own, buy, or sell land. This understanding became a grave problem when they met the Europeans. Each group learned to use each other's language and concepts, but their understandings of them were totally different. The Native peoples thought that the settlers who bought Manhattan Island were crazy. How could you own, or buy or sell, this womb, this earth mother that gave you birth and would take you back? It was incongruous. This group's difference of perception was not so much rooted in each one's religion as in their cosmology.

Now let's look for a moment at the cosmology which has shaped Western civilization, for the difference will make some sense when we look at what's happening presently on the planet. Fundamentally, we're at a point in time where we're all shifting our perspectives about cosmology. We have new insights, with new implications. But we're living within institutions that are totally rooted in the old one. And they're inflexible. You can't make a shift within structures as easily as you can make a shift in understanding. So the ambiguity and the tension and the conflict we're experiencing is a part of the process, and there is no small pain involved in dealing with it.

Basically, the premises within Western cultures are derived from different origin stories, but there are a couple of implications in those stories that have the same sense.

The first is in their perception of the Divine, or the deities, or the gods, however they were imaged. They were seen as being fundamentally transcendent to the universe; they lived outside of the universe; they were greater than and different from the universe. God was a transcendent being who had power and dominion over it, and was far greater than and different from it. Apart from it . . . other than it.

Secondly, there is a sense of the human as being intrinsically connected to the Divine, as having a major significance, being in union with the Divine. But in order for the human to do that, the human, too, had to transcend the world of material reality to go to the ideal world. So the locus, the meeting place, was transcendent.

Finally, this left a perception of the world in purely material terms. Because the world was not involved in that process, it didn't have an inherent spiritual dimension. It was understood in totally material terms. It was a reflection of the divine, and was therefore holy, but it was not spirit, it was material. This perception enabled Western peoples to sense they were detached from the world, and possessed dominion over it. In some instances that was terribly abused, in other instances it evolved into a sense of stewardship. Stewardship was an ethic of caring for what was a reflection of the Divine. But, the world was still material.

Ironically, this sense of detachment is what enabled us to probe the world. We were observers, so therefore we could explore the world and figure out how it worked. We could look into its physical, purely material, mechanistic parts and figure out how they worked. We could use that understanding as applied technology. Whether it was as simple as creating a wheel, or whether it was the kind of technology we're shaping today, we can do what we're doing only because of the cosmology.

The whole evolution of Western history is marked by this ever-broadening and deepening knowledge of the physical energies of the universe and their application in technology. It was the cosmology which enabled us to take what we thought was the last dense piece of matter—the atom—in our century, and split it open. Of course, in that event, we came to the realization that our cosmology was completely inadequate. We realized that we were living out of a set of assumptions that no longer could underpin our world or activities, because they were no longer truthful.

As we have probed the interior of the atom, we've come to see that its inner dimension is not material. It is not measurable; it has a deep spiritual, psychic, inner dimension. Likewise, as we've probed the outer universe, as we've learned its age and its unfolding nature, we've shifted our fundamental assumptions about both the universe and the earth. We've had to move from this ready-made, totally furnished, spatial universe, which we simply inherited to exist upon, to an understanding of the universe that is itself in process, which from the very beginning has had a deep, *spiritual interior*. This interior aspect, too, we realize has been expanding and unfolding over eons of time. Our planet and our solar system are recent manifestations, recent developments within a sequence of events that began fifteen billion years ago. We have a direct connection with that. It defines who we are, our crises, and our way of opening up into the future.

At this point of our human journey, we're just beginning to grapple with a universe that has a fifteen-billion-year history, from its first emergence as hydrogen out of the mystery which brought it into being. Now, I happen to be a person of faith, and so my explanation of that origin is that it comes from a Divine source which has within it all the potential that the universe can possibly express. But whatever our understanding of its beginning, we are the first generation on the planet, at least, to have the story in a new context of expanding time and space.

So, if we were to look at that fifteen-billion-year process, from the beginning until now, we'd see that the universe began with hydrogen. But out of the hydrogen atoms came helium. And out of the unfolding of helium and hydrogen came carbon, all those differentiated elements by which the universe unfolds to greater and greater levels of complexity. These elements have the capacity to unite, make new combinations, and become ever more complex. From the very beginning, as we're discovering, there are spaces within the hydrogen so vast that they're immeasurable. Thus their interiority develops as well. Not only does the universe begin to unfold externally, but its interior dimension unfolds, evolves to a high psychic complexity in order to realize its inner potential.

When our sun came into being some five billion years ago with our solar system, atomic elements were formed which became the heavy metals eventually forming the crust of the earth. There are the unique elemental structures within the atoms of the sun and our planet. But this is a single continuum; this is one event, one event present to itself in the unfolding of the process. If we look at our earth over the past five billion years, we begin to see an extraordinary acceleration of that complex structure. But it's the universe we are describing in the creative process of the earth.

Now, I can't imagine five billion years; that number totally escapes me. You might as well say five million as five billion. Some popular scientific writers help us to image this time frame by compressing five billion years into a twelve month period, watching the process unfolding in a stepped-up speed. So let's do that. If five billion years could equal one year, then we could begin to see how the earth developed in its potential for life.

From its first gaseous state, to the formation of life, it took the earth about eight months, the first eight months. The higher capacities, such as breathing, sensory capabilities, reproduction, and self-healing, took place in the last four months. But it's the earth as a subject performing these functions. It's the earth, through its inner psychic dimension, that is acting in these new sensitivities.

It would have taken the last four months of increasing complexity and diversification for the earth to unfold within itself a brain so highly evolved, a nervous system so highly organized, a skeletal structure so highly developed, that the earth became capable not only of living and breathing, moving, feeding, reproducing itself, of seeing and hearing, but now it had evolved an organism so complex that the earth became capable of thinking about itself. And that's the human.

The human is the being in whom the earth has become spiritually aware, has awakened into consciousness, has become self-aware and self-reflecting. In the human, the earth begins to reflect on itself. In our deepest definition and deepest subjectivity, we humans *are* the earth. Conscious. You and I are the beings in whom the earth thinks . . . knows . . . comprehends . . . analyzes . . . rationalizes . . . judges . . . remembers . . . chooses . . . acts . . . decides. Of course, we're not used to thinking of ourselves in this way at all. As a matter of fact, it's rather upsetting. We don't know what to do with it. It doesn't fit into any of our categories. As Teilhard de Chardin said, the human person is fifteen billion years of unbroken evolution now thinking about itself. That's who you are. And you are irreplaceable and unrepeatable. The way the earth is thinking at this moment in you is *unique*. Totally. And it will never think that way again.

Let's look at the consequences of this. In this twelve-month time frame, the human, too, has been around for only one day. We're one of the youngest species, very primitive. It might be a miscalculation to say we're human; we might still be pre-human, except for some great enlightened beings who walked the planet as more fully human. But we've only been here one day, and we're very young! Now if you could take this last day and look at the twenty-four hours that we've been around, we know that the majority of that time was spent in that tribal age of which we know very little. This is where consciousness began to awaken and unfold, and our human ancestors went through the extraordinary process of learning to abstract thought, shape symbols, and create something as highly complex as language, to learn to survive in extraordinary circumstances while developing the social systems and the myths and stories which became the basis of the earliest stage of culture. But, the later civilizations which we describe as ancient history only began about thirty minutes ago. And it was in this time frame that we wrote our cosmology, our earliest science, our earliest explanation for how the universe came into being.

So you see our cosmology is relatively new. For thousands of years, it has given great coherence to our sense of purpose and meaning. It was a paradigm which provided a story of how we humans fit in the universe. This story has provided the meaning which has guided us into the present. It was not evil or wrong. It was just our explanation of reality. It was what we knew. But what's happened in the past half-hour and especially in the past few moments, in what we might call the scientific-industrial age, is that our knowledge has exploded, and our power has expanded so astronomically, that through us, and what we now know, the earth is coming to a new moment of awareness. It's learning its own story. It's learning where it came from and who it is. It's coming into a new phase. And we, who are the five and a half billion people around the rim of that little planet, we who are its thoughts, are the ones in whom, right now, the earth is coming to awareness. Our understanding has become so deepened and broadened that we are literally bringing the earth into a whole new phase in its unfolding. This is as radical as the shift from non-life to life, or life to conscious life. The shift that's happening now is that consciousness has such a knowledge and understanding and power over its own process. It is starting to take control of the process. It's going off "remote control."

The past process of the universe and earth has come about through an "internal guidance system," for want of a better term. Now we, and therefore the earth, know how that system has been operating. And we have the power to turn it off "automatic" and put it on "manual." That's what's happening. Consciousness is taking over. Now that's extraordinary. We could all walk out of here right now and just deal with that. But I want to make that specific so that we can see the depth of what it is that we're talking about.

Look at the area of genetics, for instance. Since we've broken the code of DNA, we've gone inside the chromosome; we know how it works, we can figure out the genetic memory that has been accumulating and is now imprinting itself within a living cell though this whole time frame. Consciousness understands it. And our knowledge of it, our power over it, is giving us the capacity to go in and rearrange it. We can interfere with the natural process and alter it to whatever we decide we want it to be. That's what's happening. So in a very real sense, the breakthrough in bio-genetic engineering is the story of the earth coming off "remote control" and beginning to consciously shape the future evolution of its life.

And, of course, the question for us, who are this consciousness, is have we arrived at anywhere near the level of integrity, wisdom, or maturity to do what we can do with the truthfulness that life had when it was on "remote control"?

We're like adolescents. We have these extraordinary powers, but without the life experience to integrate the power into a large context. And the question is so profound because presently, we're preparing to release into the web of life organisms which did not pre-exist, which did not come out of the slow, laborious process of evolution, where at each stage all organisms worked together in a finely tuned balancing act to enable the conditions for the survival and ongoing process to continue toward life. If we err, if the finely tuned balancing act which has sustained this planet and helped it develop into the most extraordinary life community in our solar system, if this gets violated, if it goes off, or becomes totally whacked out, the only cause of that will have been its own consciousness. Without consciousness, while still on "internal control," it was moving toward life, and toward *sustainable* life.

And it's not even so much a question of ethics, because you have very fine, righteous people involved in biotechnology. They love their husbands or wives and their kids. They don't tell lies. They're pure. They're good people. Their goodness isn't the issue if they're operating out of the old cosmology. This isn't a question of ethics. It can *become* a question of ethics. But you see, we don't have a tradition of ethics out of which to judge this. Our cosmology created an ethical system that was human/human, or human/God, but not human/universe process. We don't have it. The indigenous peoples do, but Westerners do not. We do not relate to the earth in this ethical manner.

Here's another example. We now know that we're alive because the earth is alive. Unlike Mars, or the moon, or Venus, or the other planets in our solar system, we're a water planet. Seventy percent of the earth's surface is salt water. That's why the earth is alive. Its a fluid planet. But in our old cosmology, we call these fluids oceans. We name them . . . Atlantic, Pacific, Arctic, Antarctic. They're places. They're things. They're *its*. They're *stuffs*. You swim in them, you fish in them, you sail in them, you own them. You own homefronts on them. And if your cosmology is such that those are just places, then it's very logical to dump wastes there, including our very lethal wastes.

But now we're beginning to understand that the oceans are the actual fluids of the planet. And everything that lives has the ocean in it. The oceans are not oceans. They are one single salt water system which flows through everything on the surface of the earth that has life in it. That's why things are alive—maple trees, bananas, or you. If we took you to the chemistry lab and had you analyzed right now, regardless of your size or weight, you would be seventy percent salt water. And it's the same salt water as if flowing through the oceans. The rest of you would be the minerals that form the crust of the earth. *We're the earth, with consciousness, with soul, with spirit.* We're the earth in a new form. But we are the earth!

And now we understand that these fluids within the oceans are in us. Because the oceans become the clouds and the clouds become the rain and the rain becomes the corn. And we eat the corn. And we get our minerals and our salt water replaced. And we cry the ocean. We excrete the ocean.

We are just beginning to realize that the oceans are alive because over this long, painstaking process toward life, they became a community of millions of varied species and organisms, all of which are a fabric and a community of life. They are totally interdependent, all essential for each other's existence and for the well-being of the whole earth so that it can function and constantly maintain the oxygen needed by everything that lives.

As we continue to dump our toxins in the oceans, we're beginning to see gaps in the fabric. These marine organisms never evolved with the capacity to endure this sudden onslaught of poisonous new substances. Many can't reproduce. They're becoming extinct. And as one becomes extinct, the food chain gets altered. And as those toxins build up in the food chain, more complex species are becoming extinct, so that the oceans could literally die. Jacques Cousteau says we have very few years left. If we don't change what we're doing, the oceans are going to become toxic; they will have lost their capacity to break down toxins and to keep oxygen flowing. If the oceans do become toxic, then the clouds are going to be toxic, and the rain will be toxic, and the corn will be toxic. And our children will be toxic, and their tears will be toxic.

If the oceans die, that's literally the death of the planet. And if the planet dies, the only cause of it will have been *consciousness*, because without *consciousness*, the whole thing was coded toward life. Something's interfering with the process. There are dynamics happening at the most profound level which are altering the capacity of the earth to do what the universe has mandated it to do. That is to continue to live and to continue to heal and nourish and regenerate itself. Consciousness is violating this mandate. And that's us.

What binds all of us in this terrible crisis is the realization that through us, the earth could actually choose its own suicide, its own self-destruction. Basically, this is what we're talking about. Because our consciousness is so young and primitive, we don't understand the magnitude of our behavior. We don't have a context adequate for what we know and what we can do. We're still unable to understand who we even are as humans in the community of life.

We're still operating out of old assumptions and patterns that deal with war and conflict in totally inappropriate ways. Totally immature ways. But that pattern of behavior is now coupled with our knowledge of splitting the atom. We're fooling around with radioactive isotopes in our weapons, not cannon balls. So, the fifty thousand and more nuclear weapons that we have buried under the skin of the planet, under this living tissue, into this live organism, these are literally tumors.* If they go off, they are going to spread. And like tumors, they will make the whole life fabric go awry. Atoms do not know how to behave when they've been affected by a radioactive isotope because that isotope is unbalanced, and it hits on the atom of the one next to it, and knocks its whole genetic memory out. And there's no way to contain

*Editor's Note: Since this talk was presented in 1986, the number of nuclear warheads has decreased significantly, but the potential for total destruction is just as real today as then.

this. It knocks out genetic memory! We've got fifty thousand tumors in the Mother, and the only place they've come from is our *imagination*. An imagination at the service of a perception and understanding which are totally inadequate for what we, in fact, know how to do.

Each of these "nuclear tumors" is wired into a nervous system of computers and satellites and radar. If one goes off, just like in the body, it's almost impossible to contain the process. We're describing the planet's capacity to choose its own death. And if this happens, it's because of consciousness, because without our human species, this could not happen. Its a conscious choice. And what's the consciousness? The consciousness is the five and a half billion of us. The way you and I think is not irrelevant. We cannot be neutral. We're totally relevant and totally significant. It is our very consciousness which is making the decision and doing the probing. How I think affects the whole. The earth thinks as you think. The earth thinks as all of us think. And the earth is in a process of coming out of its adolescent fixation with itself and its powers, into a whole new level of maturity. And to the degree that you and I make that jump, the earth makes the jump. It's as simple and as profound as that.

Ironically, it's the physicists among us who seem to be most in touch with this as they explore the inner dimensions of space. They are touching and probing and contemplating interior realms of energy and of activity that are beyond our comprehension. It's the physicists who are beginning to speak in metaphysical language with new ethical and theological insights. It's the astronauts, too, who are becoming the mystics of this age. They are being changed by their new perspectives of inner and outer space. This is not because they happen to be morally better people, but because they've got a different view. They've got a whole other perspective into the nature of creation and it's causing a humility that is, to a certain extent, unprecedented.

It's a new revelation.

Now, let's consider certain principles pertaining to why the universe has unfolded, and why our particular little planet is so resplendent with life. We now understand these principles through an empirical observation of different levels of reality. What is so astounding is that these principles do not deny or contradict the deep insights of all the earth's spiritual and moral traditions, but rather expand and complement them. The principles are extremely simple, and of course, that's why they're so difficult. What I'd like to do is try to put them into a perspective. They're interconnected. You can't separate them from each other. They don't exist except that they go together. And they underpin everything.

The first is the principle of differentiation. Simply put, it means that the universe works because it's coded to differentiate itself toward greater levels of complexity. So, if the universe had stayed just hydrogen, it would still be hydrogen. You can only have a universe because it's coded to differentiation. Now that's an essential, integral principle. It implicates everything. So, once there is differentiation, there can be hydrogen and helium and carbon, and so on, and because there are these differences, they interact with and change each other. But, don't forget. It's all connected. The universe, even in its simplest forms, is a single event with itself. It's present to itself, not only in its external, but in its inner dimension. It's a communion. Because you have different elements, they come up against each other and cause each other to change. The changing process causes what is simple to break down. When it breaks down and interacts with the other, what is released from within itself is the potential to become more. This is a fundamental law. This is what enables complex elements to evolve. And when our earth comes into being, this process simply takes off. The earth has developed so highly, has become capable of life and of consciousness because of this particular principle.

Let's look at the earth's crust. It's essential that all those differing elements be there. Our chemistry chart describes those elements as the basic component parts of everything that is on the planet. You couldn't have a living earth made of just iron, or just calcium. It wouldn't work. It is all of those elements that become the substance of the planet and then unfold into simple organisms, and then into more complex ones. At every stage, we can see the conditions by which the whole system is working. All those differing organisms unlock the potential still within the process to become transformed into something more complex, more capable of life and consciousness. It's the law. There couldn't be a green planet if the only vegetation that existed were maple trees. It couldn't work. If the only insects on the planet were fleas, the planet would have died long ago. Within a shovel full of soil are millions of organisms and microbes, which, by the way, are the real farmers. They're the ones that produce food. Farmers don't produce food. Part of our problem is that farmers think they produce food. No. Microbes grow food. That's why agriculture is collapsing; the microbes are gone. It's just a matter of time.

So, because those essential elements are all there, this total diversification within the planet enables the next stage to unfold. That's coded in, and it progresses. And the genetic coding within a particular life form continues into the future and is carried forth into the next stage. When the earth finally becomes conscious within the human, differentiation enters another level, even within its physical diversity. There is black and white and brown and red. Those differences, those racial differences and characteristics within particular people from particular geographic areas are as much coded in as the difference in oak and birch and sycamore and pine. It's the truth. They're not mistakes, they're not errors, they're not miscalculations. It's the truth. There are male and female. That's the truth. And from the moment that the universe began to reproduce itself, to regenerate life sexually, in the most primitive organisms, from that moment on the universe became sexual and the differences between male and female became the process. That's the truth.

Now if you believe in a Divine Creator, then this sexual condition for creation or regeneration is a reflection of the way the creator designs it. This is truly revelation. If the universe evolves in this way, then this process reveals the mind of the Creator, or the origin principle. We're just starting to catch on to this. But some people haven't. If their only frame of reference for revelation is through human channeling and the revelation was channeled out of an old cosmology, well, they're not going to find this revelation there. Because it wouldn't be perceived. You know what I'm saying? The truth of the universe itself is prior to what we call revelation. The differences in the universe are the truth. So then, how people consciously become aware of reality, reflect on it, and make their judgments and decisions, is also rooted in this impulse towards differentiation.

Just as wrens or nightingales, robins or thrushes, are all genetically coded to a particular song, but sing differently, so humans participate in the shaping of their language. All humans are coded to speech. But because we create the symbols of speech, the languages of the earth's peoples are different. That is not a problem.

All peoples create architecture. But they create architecture out of reflecting on the experience of the world around them and of the materials at hand. And though we all might create temple architecture, the temple architecture of the earth's peoples is going to be differentiated. And that's the truth.

We all mourn our dead but we sing our dirges in different languages. Yet, the dirge itself is universal. And we all reflect on reality and try to grasp and be open to the deep mysteries. But Jew isn't Hindu, and Hindu isn't Buddhist, and Buddhist isn't Aztec, and Aztec isn't Christian, and Christian isn't Mayan, and Mayan isn't Dakota Sioux. And that's the truth.

And all humans take their meaning systems and connect them with their history and their experience and their culture and their language, and their wisdom and their traditions, and they shape culture. But Polish isn't Greek. And Greek isn't Russian. And Russian isn't Eskimo. And Eskimo isn't Mayan. And that isn't Czechoslovakian. And that isn't English. And that's the truth.

Now the problem is, you see, that unlike the birch and the sycamore and the oak and the maple, we humans reflect on ourselves, and we can say, "I." "I am." Once we can say "I am," we say, "I am true." Once we claim the truth of our being, then we experience a certain susceptibility, a certain inclination to reflect on our own truth, as being in the image of the "Big Truth," out of which we've come. That's still all right. But when we see somebody come along who's different, we have this tendency to say, "I am the truth. You are different. You obviously cannot be the truth. So since you're not the truth, and I am the truth, then it's really an imperative on my part to make you conform to my truth. So you'll be all right. And if you don't want to do that, then you obviously are not the truth. And since you're not the truth, and therefore not fully human, then it's totally legitimate to enslave you or to reduce you to another level of definition which is not truth. And if you don't like that, we'll just annihilate you."

There's some strange tendency in the human stemming from our self-awareness that wants to force others to conform to our criteria of truth. And we have this love-hate relationship with the tension between truth and conformity. Individuality and conformity. It's a thing we have not finely tuned yet, perhaps because we're still so new to it. But in the last half-hour, we've recorded that we've fought over eight thousand wars with each other. I suspect part of the cause of those wars is this tendency to demand others to conform to a perception of truth as we perceive it. It has not yet occurred to us that the capacity of the human to reflect on the truth *must bring about differentiation*. And rather than the differentiation being a problem to solve, it's a solution to the challenges of any existence.

Just as you cannot have a living planet except with all the diversity of minerals and vegetation, so it is within the realm of consciousness. No single consciousness can totally reflect what is infinite. Ironically, it's the physicists who are warning that we'd better learn that or in spite of our being "good people" who love our kids and don't tell lies or do unethical things, we will continue to violate the very process out of which the universe has been coded by its creator. I think they are trying to help us to understand that the human is the being in whom the universe can finally *comprehend* the need for differences. We must grasp the differences. We have the capacity to contemplate the differences and to delight in them and celebrate them. That's the proper mode of the human! That's who we really are! We're the ones capable of delighting in the designs that came out of the process.

A second principle of the unfolding universe is that of interiority, which simply means that things are different because everything that is, is itself a truth. It has its own inner reality which makes it itself and not anything else. Hydrogen is not helium. Mary is not Fred. Fred is not Gary. Black is not white. Male is not female. And whether you're speaking of the atom in its simple substance, or whether you're speaking about a Mozart composition, *this is this*, each is what it is. And therefore, it's different. When we probe the interior spaces of things, we begin to understand that we go beyond what is measurable. Our physicists are coming in touch with the fact that what's at these deep interior levels is being generated as they're observing them. This means these energy patterns are coming into existence and going out of existence. They're being created in the moment. The center of any one interior is the center of all interiors. The volume of any one atom is the volume of all atoms. At its depths, it is not material. Therefore, what has come to be expressed as hydrogen, or Mozart, is a revelation of this and it's truth.

This is why the universe has expanded. In human beings, the universe can comprehend this process, and out of our capacity to see the interior of the other, to see its truth. We're capable of reverence. We're capable of non-violence toward the uniqueness of the other. We're either going to learn this quickly or we're going to die.

In this century alone, we will have caused the extinction of close to a million species of plant and animal life. And not even God brings back extinct species. To say nothing of the individuals and cultures that have simply been slaughtered because of their differentiation or because we didn't understand their interiority. Some groups of people with their unique culture are being pushed toward annihilation. This is totally against the laws of the universe. If their wisdom is lost, it will never be regained. The child silenced is silenced forever, and that voice will never be a revelation of the Creator that it was meant to be. Our problem is we don't understand revelation.

A third principle is that of communion, which simply means that from the very beginning the universe has been in communion with itself; that no differentiated interiority can exist alone. It can exist only because it's a member of a bonded community that has evolved in a mutually interdependent way with itself. The universe indwells with itself. There are no empty spaces; there are no vacuums; there are no islands.

So the earth has evolved as a single communion with itself, both inwardly in its psychic development and externally in the whole composition of its diverse forms. It's a single organism. If you could see into my hand you would see every atom articulated, every atom different. But every atom is bonded so that my hand operates as a whole. The universe and the earth do not violate these principles. Communion does not demand conformity. Conformity and community are diametrically opposed. Uniformity is a violation. We're beginning to understand that the human is the being in whom this can be understood with awareness and freedom. This is the meaning of love. Love is the bonding of the planet through awareness and through freedom. This is our true human destiny, to be entranced with the whole; to love it and become the ground of its being.

And love is not love when it demands the conformity of the other. As we probe the depths of the inner space of the human psyche, we're starting to understand that. In other words, if you know who you are, if you know who your interiority is, your unique manifestation of life, if you're in touch with that, and you're aware of who you are, and you're affirmed in that and you know you are good and lovable, then you are not threatened by me if I'm different. If you know who you are, you have the capacity to delight in my differences. And the more you delight in me, the more I can dig into fifteen billion years of energy and keep pulling out all kinds of treasures. If you affirm me because I have different tastes than you, I'll develop that. I'll become a poet. You affirm me, I'll become a playwright, I'll become a musician, I'll become a computer analyst. You become the ground of my being, and the more you affirm, the more I'll bring out. I will even become, at your rejoicing, capable of compassion, and mercy, and gentleness, and justice, and integrity, and peace. So that you bring out of me, you activate in me the deepest mysteries of the universe. That's love. And you can do it without liking me! And it's this the physicists are saying to us. We've got to learn it, it's got to become internalized, and it's got to become the inner core of our institutions and culture.

You know, when you put these principles together with the discoveries in human intuition over the past half-hour, you begin to see that we were shaping a way to touch the deep inner mysteries of existence. Our religious stories express how we were grappling with the deep force of the universe all along! But we didn't have the evidence available to see as we do now. We were attuning to the medium of the spirit.

We can trace this through any of the religious traditions. But to reflect on my own, which is the Judeo-Christian, is to open up new understandings. For instance, when Israel as a people evolved, when *through Israel*, the earth came to a new moment of consciousness, they conceived of God as one, not many, as faithful, not arbitrary. God was simple and loving and life-giving. When such a consciousness evolved in the Hebrews, then that was a moment of breakthrough. Israel understood that it was to participate with that life-giving God in a very real way and to bring that energy and that spirit into time and into history. So, if Israel was faithful as God's creature, adhered to a covenant relationship with the Creator, if she was obedient, then though her life and the work of her hands, she was bringing the energy of God into time and place.

It was based on the notion of a covenant. She had to be obedient to the law as revealed through Moses in the Ten Commandments. That law is really an articulation of the three aforementioned principles of the universe.

Likewise, Jesus carried the same Hebrew consciousness and expanded it to another realization. He reduced the commandments to two. The first one is, be conscious in your humanness of who you are; be conscious with your whole heart, mind, soul, and strength; adhere to the Creator in whose image you're made. Then out of that, hold everything in communion. Love it as yourself. His breakthrough in consciousness was to be aware of and choose the communion of everything.

One of the most significant realizations within the process of the universe is the vital dynamic of *change*. Because these principles are functioning, there has to be change. The universe cannot remain static. As soon as there are different components interacting on one another, each has to break down and change. They enter into chaos. They have to let go of being what they were, and in that chaos, that breaking down, they breakthrough to a new level of being. Thus, there is the event of energy, of breakdown, and of breakthrough.

When that dynamic emerges within the realm of human consciousness, it becomes the life, death, and resurrection cycle which is at the heart of all of our spiritual traditions. This dynamic plays itself out in a unique way within consciousness. It means that when the human deals with some one or some thing that is different, it calls for a new exercise of power, not power over the other, or a power that causes conformity or annihilation. Rather it is a power to enter into a relationship of dialogue or communion, wherein chaos is experienced. Both pass into change. And then out of that comes a deeper wisdom which is a transformation.

This life, death, and resurrection is out of levels of fear and ignorance and prejudice and hostility. It brings about the clarification of the deeper self and it is totally congruent with the health of the universe. As a matter of fact, it is the way to choose life. Otherwise, we choose death. Personal and collective.

Let me end finally with two implications. This new cosmology, which is our connection to a fifteen-billion-year process, causes us to do two things. One is to come home. We've got to acknowledge our identification with the earth and with the spiritual dynamics of the universe. We are literally stars thinking about themselves. This is the context of our true existence. We can't have our life, or nourishment, or learning except as it comes out of the earth, which is our very body.

And the earth as a body is a communion and a community. At its deepest level, it's in communion with itself, and in its external manifestation, it is a community of all the beings who share existence on this planet, from atoms of oxygen to the most complex organisms. And if this earth is sick, we will be sick. We have to come home.

We've got to leave the dualisms of the past and become members of the community of home. I mean home now in terms of the literal sense of where we actually live on this planet. We live in a particular region. Long before humans ever came into it, this region was marked by a unique geological history, a unique temperature, unique water systems, unique climates, and macroclimates. These were the actual conditions that evolved over time, and determined the kinds of vegetation that could exist there. Over long eons of time, a unique community of vegetation worked out a balancing act wherein all could coexist in total obedience to a principle of differentiation, interiority, and communion.

All species that came into the community had to abide by these laws or they got kicked out fast. They went into oblivion. Only the ones that could cooperate in this balancing act survived. Only the fittest survived—those who could fit in with the whole. So, the insects, birds, fish, and animals who learned to cooperate became the integral community of life. Unless we learn that this community is the source of our health and our nourishment and our governance, our human cultures are not going to make it.

We've got to come home to the prior community because the community is sick. Our water is sick, our air is sick, our soil is sick, our vegetation is sick, and they no longer even inspire our souls, our poetry, our children, or our understanding of God. Thomas Berry asks the question, "How will we baptize our children with toxic water and tell them about God? How will we give pills and medication out with toxic water, and hope to make people well? How will we eat contaminated food, and think that we will be nourished?"

We've got to come home in a new humility. And become members of the community, or else we're simply going to be a bad experiment on the planet, and get kicked out.

Secondly, we've got ourselves organized into a hundred-and-fifty-some-odd nation states, and we still think that the best way each nation state can secure its survival is to compete against the other hundred-and-fifty-odd nation states. And so we continue to mobilize all our internal resources to produce the nuclear tumors and to compete against each other. And that very competition is what is causing the death of the air and the soil and the water and the whole bit.

For the first time, we're in a new place. We have a new revelation. We are literally the first humans to go off the planet, to go out into space, turn back, and look in the mirror. With the image of the earth in space, we have a new image of who we are and what our destiny is: the beauty of it, the magnitude of it, and the silence in which it finds itself!

Let me end finally with a favorite reading that I have on hope. It comes from a Brazilian theologian by the name of Ruben Alvez, who wrote "Tomorrow's Child." He said:

> What is hope? It is the pre-sentiment that imagination is more real, and reality less real than it looks. It is the suspicion that the overwhelming brutality of facts that oppress us and repress us is not the last word. It is the hunch that reality is more complex than the realists want us to believe. That the frontiers of the possible are not determined by the limits of the actual. And that in a miraculous and unexpected way, life is preparing the creative events which will open the way to freedom and to resurrection. But, the two, suffering and hope, must live from each other. Suffering without hope produces resentment and despair. But hope without suffering creates illusions, naiveté, and drunkenness. So let us plant dates, even though we who plant them will never eat them. We must live by the love of what we will never see. This is the secret of discipline. It is a refusal to let our creative act be dissolved away by our own

need for immediate sense experience. And it's a stubborn commitment to the future of our grandchildren. Such disciplined love is what has given saints, revolutionaries, and martyrs the courage to die for the future they envisage. They make their own bodies the seed of their own highest hopes.

—27—
Creating an Ecological Economy

Petra Kelly

Is this not a precious home for us Earthlings? Is this not worth our love?
Does it not deserve all the inventiveness and courage and generosity of which
we are capable to preserve it from degradation and destruction and, by doing
so, to secure our own survival?

—Barbara Ward and Rene Dubos[1]

Ecologist Aldo Leopold had an experience in which he began "thinking like a mountain." He came to see humans as "plain members" of the natural world, writing, "We abuse land because we regard it as a commodity belonging to us. When we see land as a community to which we belong, we may begin to use it with love and respect."[2]

Like the mind-set that places men above women, whites above blacks, and rich above poor, the mentality that places humans above nature is dysfunctional. It is based on the principle of domination. We humans take our earth for granted and never hesitate to exploit it and its other inhabitants to gratify our immediate wants. We have to understand that we are part of nature, not outside of it. What we do to the earth, we do to ourselves. Understanding our interconnectedness with all life is the essence of ecological politics and an ecological economy.

We need little of what consumer society tells us. Lavish consumption brings only a crude sort of gratification. Cultivating the intuitions of our identity with the whole life in all its diversity brings delight to the heart and a deeper and more durable fulfillment. These intuitions are essential for a life based on values, and they are also the basis for political action. The personal transformation to what Arne Naess calls "a life simple in means but rich in ends" is itself a political act. If we want to transform society in an ecological way, we must profoundly transform ourselves.

E. F. Schumacher said that a nonviolent and gentle attitude toward nature must be the basis for all politics:

> The violent and aggressive approach to the natural world is fed by human greed for short-term material gain without care of the long-term ill effects on other generations. Anyone analyzing Western economics will realize very soon how violent and aggressive the

Western approach has been up to now in this area. Western economic thinking depends upon insatiable consumption—demanding more and more and larger and larger goods and services to be available at all times. But where is the basic ethic of restraint?

In telling us that "small is beautiful," Schumacher points to a consciousness of the limits in which we must live in order not to degrade our environment and ourselves.[4]

In fact, we have already far exceeded those limits, and we continue to do so. Yet no official economic policy has taken seriously the accelerating biological holocaust devastating the earth. All of the established parties refuse to recognize the need to curtail economic growth and not promote it if we are to avert a complete ecological catastrophe. Despite all of the environmental evidence that their way of doing business is destroying life, governments continue to make decisions based on short-term economic gain. In the end, their economies will pull the rugs out from under their own feet and bring the rest of us down with them. The earth simply cannot sustain unlimited exploitation.

We in the Greens say that we have borrowed the earth from our children. Green politics is about having just "enough" and not "more," and this runs counter to all of the economic assumptions of industrial society. We must question those assumptions. The ecological crisis is a crisis of consumption, not of scarcity of resources. The industrialized countries must move from growth-oriented to sustainable economies, with conservation replacing consumption as the driving force. In the Third World, economic and ecological development must be addressed together if we wish to achieve just and sustainable results. But this cannot happen if the North's exploitative policies continue. Ecologically balanced development depends on a just redistribution of wealth from the North to the South. Environmental problems cannot be addressed apart from the economic issues they are linked with. We have to ask, How would an ecological economy function, and what can we do to bring it about?

An ecological economy would measure the prosperity of a society not in terms of the numbers of goods produced, but rather in terms of production methods that conserve the environment, protect human health, and result in durable consumer goods. The measure of value would include clean air, pure water, unpoisoned food, and the flourishing of diverse life forms. In an ecological society, the economy, lifestyles, and consumer expectations would be characterized by considerations of human and environmental health. The economy would not regard industrial growth as its guiding value, but would be guided by respect for life and the inherent worth of nature. The relationship between humans and nature would not be a one-way exploitative process, but a partnership based on interdependence.

The establishment of an ecological economy would require the partial dismantling and conversion of existing industrial systems, in particular those branches of industry that are hazardous to life, above all the nuclear, chemical, and defense industries. Wherever possible and ecologically meaningful, mass industry would be replaced by small, decentralized production units that are sparing in their use of natural resources and produce a minimum of hazardous waste. Decentralized production can be more ecologically sound, more responsive to the needs of the workers and the immediate community, and the basis for greater local economic autonomy.

Ecological production would reject the sharp national and international division of labor that has led to large economic imbalances between regions, between city and countryside, and between industrialized nations and the Third World, and has brought about high transportation costs and correspondingly excessive energy and land requirements. Locating production facilities closer to consumers would reduce transportation costs and energy consumption. These changes are within our means. Their implementation depends only on our will but will probably not take place without grassroots organizing, because national political leaders cannot see or refuse to recognize the limits of growth, and they lack the vision or the motivation to base their decisions on environmental ethics.

The changes just stated apply only to economies that are already industrialized. What about the developing countries and the traditional and aboriginal societies? Over the years, I have seen many disagreements between the environmental movements in the Western affluent societies whose focus is solely on ecological concerns, and activists in the Third and Fourth Worlds and from oppressed communities within the affluent countries who insist that ecological concerns must be tied to issues of economic justice—the exploitation of the poor by the rich. The rape of the earth will not be halted simply by the affluent imposing conservation measures on others. The seedbed of ecological destruction is the global division between the rich and the poor. Behind every environmental issue are landless villagers or workers who are forced to destroy nature in order to survive, or governments, banks, or corporations who pursue economic growth without much regard for people or nature. Ben Jackson wrote, "A series of 'Keep Out' signs around the world's forests would not only be morally unacceptable, but with the world's poor still hungry outside, it just would not work." To solve our environmental problems, we must address these economic issues. And conversely, to tackle the grave problems of poverty, we must nurture the environment. To quote Ben Jackson again:

> The environment directly underpins the long-term viability and
> security of many poor people's livelihoods and culture. Only if

soils are cared for, will they keep producing crops or withstand
the shocks of natural disasters like drought. The environment is
also the medium through which many of their essential needs are
met or not met. Health, for example, is often dependent on
whether the river or well gives clean drinking water and whether
there is decent sanitation.[5]

The transformation of forests into deserts, fertile earth into sunbaked
concrete, and running rivers into silted floodwaters show that only through
care for the environment can the livelihoods of those most dependent on it
be sustained. We cannot allow economic and environmental concerns to be
played off against each other. We see this over and over as corporations and
governments portray the environmental movement as undermining and dis-
regarding the welfare of humans. It is incumbent upon the environmental
movement to find common ground and establish new alliances with groups
concerned with peace and social justice, world poverty, and labor issues.
These issues are all linked. We can take inspiration from Australia's Green
Ban, in which trade unionists refused to work on environmentally damaging
projects, even at the cost of jobs and imprisonment.

The lifestyles, policies, and production methods of the industrial countries endanger human existence and can only be maintained through the
increasing exploitations of the Third World. An industrial system predicated
on the delusion of limitless expansion will, in time, consume its own basis of
support. Obviously, such a system cannot be the model for development for
the Third World. Ecological development is possible, but we must insist on
changing "business as usual."

Poor countries continue to finance rich ones on an enormous scale. Between
1982 and 1987, Third World countries provided a net of more than $220 billion
to the wealthy countries, mostly in necessary imports and interest payments for
so-called development loans. The debt figure is now more than $1 trillion. In
1989 alone, the World Bank took $724 million more out of the Brazilian econ-
omy that it put in. Since taking the World Bank's bitter medicine in 1983,
Ghana's debt burden has doubled to an estimated $3 billion. In September
1989, a hearing of the Human Rights Caucus of the U.S. Congress found that
more than 1.5 million people are currently being displaced by World Bank projects. Environmentally harmful development projects funded by the World
Bank are routinely dropped in the laps of Third World communities without
any consultation. While President Bush praised the World Bank and the
International Monetary Fund as "paradigms of international cooperation," the
truth was stated more accurately in *The New Internationalist*: "The World Bank
has turned out to be an ecological Frankenstein, armed with a chainsaw."[6]

Profits from the labor of people and the trade in natural resources have, for decades, flowed from South to North. Amazon rainforests are disappearing not because of the needs of the local inhabitants but to supply cheap beef to Northern consumers and charcoal for smelting iron for export. Southeast Asia's forests are being decimated to supply chopsticks for the Japanese and tropical hardwood for Western markets. In order to develop, the Third World countries have to borrow from the North and then pay the interest on these loans by cashing in their natural resources. Many of the mega-projects that have helped establish the debt treadmill are themselves environmentally destructive, such as the proliferation of nuclear power plants. Indebted countries have not just borrowed from their futures, they have mortgaged them. And, as Susan George says, "Nature has put up the collateral."[7] As commodity prices fall and the debt burden spirals, the Third World is trapped in a vicious cycle of having to export more and earn less.

Industrial countries make the world poorer by consuming disproportionate amounts of the world's resources. The U.S., with five percent of the world's population, is responsible for twenty-five percent of its energy consumption. Australia's sixteen million people have the same impact on the world's resources as 650 million Africans. The wealthy nations account for 70 percent of worldwide carbon dioxide emissions, 84 percent of chlorflurocarbon production, and 90 percent of the greenhouse gas emissions, yet they arrogantly call upon the struggling nations to reduce their carbon dioxide emissions and refuse to do the same. Between 1986 and 1988, over three million tons of wastes were shipped from industrialized to developing countries in exchange for cash payments. This kind of "garbage imperialism" occurs because the poor nations are so desperate for currency that they will sacrifice their own health.

Maneka Gandhi, Indian's former Minister for the Environment, was pointing in the direction of the truth when she argued that Western governments actively promote environmental catastrophes in the Third World, yet they want to push the entire burden of ecological adjustment on the Third World. The latest trick is to speak of the need for "global solutions," a euphemistic way to shrug off their role and responsibility in creating these problems. As Vandana Shiva has said, "The North must bear the main burden and end its exploitation of both nature and the South."[8] This grossly unequal distribution of wealth and power threatens everyone. The deepening poverty of two-thirds of humanity and the degradation of the environment are not separate. An "Iron Curtain" still exists separating the North from the South. The Third World debt crisis is just a symptom of a world organized for the benefit of a privileged class that will stop at nothing to maintain its control.

For environmental solutions to be effective, economic imbalances must be redressed. We need sustainable development in the Third World that supports an ecological economic system, a more just distribution of wealth within and among nations, political reforms, and greater access to the knowledge and resources of the North. The South Commission proposed in 1981 a development strategy of self-reliance, recognizing that no country can be developed by outsiders. Development in the Third World must be for the benefit of the people of the Third World. Then the ecological viability of those countries can be maintained. An important step in this process would be the establishment of a "South Bank" as an alternative to the pressures and exploitative practices of the International Monetary Fund and the World Bank.

On the deepest level, the grave damage posed to life on earth is not a matter of East-West or North-South. The most deadly polarity is between human activities and the life-sustaining capacities of the earth. As Patricia Mische wrote, "We are at a new point in human history and in the planet's development when it has become critical to reconsider security priorities in light of new threats to life emanating from human assaults on the earth."[9]

The 1987 International Conference on the Changing Atmosphere concluded that the dangers posed by the greenhouse effect are as great as those of a nuclear war. In greenhouse scenarios, the accumulation of carbon dioxide and other atmospheric gases will trap heat and thereby warm the earth, causing polar ice caps to melt and the flooding of whole countries. Deserts would expand, resulting in mass starvation and mass migration. Every ecosystem could potentially be disrupted, with massive extinctions of plant and animal species and the support of human life. If industry and agriculture continue on their present course, these things could happen *within fifty years*.

When compared with such possibilities, notions like national security become meaningless. Wendell Berry writes, "To what point do we defend from foreign enemies a country we are destroying ourselves?"[10] Military policies are among the most destructive, wasteful contributors to environmental degradation. Worldwatch Institute has observed, "Again there is the irony that the pursuit of military might is such a costly endeavor that it drains away the resources urgently needed to protect against the environmental perils that are most likely to jeopardize our security."[11]

Militarization is in direct competition with people's needs for food, health care, and environmental protection. One half of one percent of the world's military expenditure for one year would pay for the farm equipment needed to increase food production and approach agricultural self-sufficiency in the developing world in an ecological way. The Pentagon uses as much petroleum in two and a half weeks as the entire U.S. public transport system uses in a year. There are countless examples of the enormous ecological cost of militarism.

Aboriginal peoples throughout the world have understood for millennia what science has finally come to see: that the earth and all that is in it are a living, interrelated system. As environmental issues become increasingly urgent and Western models of development become unsustainable, we will need to step out of our Eurocentrism and listen to the wisdom of traditional peoples concerning how to live on the earth. But, tragically, the very cultures to which we must turn are themselves under attack. Michael Soule observes that the same forces that threaten biological diversity are destroying whole cultures as well, and with them, their legacies of ecological knowledge.

Describing the "frenzy of exploitation" that is ravaging the earth's green mantle, Soule goes on to say:

> Never in 500 million years of terrestrial evolution has this mantle we call the biosphere been under such a savage attack. Perhaps the hardest thing to grasp is the geological and historical uniqueness of the next few decades. There simply is no precedent for what is happening to the biological fabric of this planet and there are no words to express the horror felt by those who love nature. In our lifetimes, the relentless harrying of habitats, particularly in the tropics, will reduce rain forests, reefs and savannas to vulnerable and senescent vestiges of their former grandeur and subtlety. Perhaps even more shocking than the unprecedented wave of extinction is the cessation of evolution of new species of large plants and animals—death is one thing, but an end to birth is something else. There is no escaping the conclusion that in our lifetimes, this planet will see a suspension, if not an end, to many ecological and evolutionary processes which have been uninterrupted since the beginnings of paleontological time.

We are facing a wave of extinctions that could exceed in magnitude the wave that destroyed the dinosaurs and many other species 65 million years ago. In the coming decades, thousands and, in time, millions of species will become extinct because of habitat destruction in the tropical forests. These rainforests account for seven percent of the earth's land surface, and more than 50 percent of it species. Since 1900, half the world's rainforests have been destroyed, and only 1.5 percent of their original total area is protected. Within one hundred years, that may be all that remains. Fifty acres are destroyed each minute, and an area three times the size of Switzerland disappears each year. Tropical rainforests are among the earth's most ancient ecosystems, having developed their wealth of genetic diversity under very stable, external conditions over thousands and thousands of years. They constitute a complex, seamless web in which all components are necessary to the whole. Because of this, they are extremely fragile and rarely regenerate. Once they are gone, they will be gone forever.

Rainforests are called the "lungs of the earth," because they produce a quarter of the planet's oxygen supply and absorb vast amounts of carbon dioxide. The loss of these habitats could severely disrupt the earth's climate and hasten the greenhouse effect. Rainforests contain the planet's richest gene pool. Half of all pharmaceuticals include active ingredients found in wild plants. Species, whose healing properties we have yet to even discover, would be wiped out. Agriculture depends on genetic diversity to develop new crop strains that can withstand constantly evolving disease organisms. The loss of species diversity would compromise our ability to produce the crops we need to heal and feed ourselves. The Greens have demanded that the Third World's debts be exchanged by international agreement for the guaranteed protection of the world's remaining tropical forests. A massive reforestation program, with trees selected for ecological rather than commercial value, must be undertaken. The destructive development programs that threaten the rainforests, including plantation schemes, large dams, ranching schemes, and road programs, must be phased out.

The fact of encroaching ecological catastrophe was tragically brought home to Europeans in the summer of 1988, when thousands of seals perished in the North Sea and washed ashore along Sweden's coastline. Scientists performing autopsies on the dead seals found traces of *more than a thousand* toxins in the tissue samples. The reasons are not hard to see. The North Sea is a dumping ground for all kinds of industrial waste. Each year, nearly 13,500 tons of lead, 5,600 tons of copper, plus arsenic, cadmium, mercury, and even radioactive wastes flow into the sea from the Rhine, the Meuse, and the Elbe Rivers. The sea is dotted with 4,000 wells and 150 drilling platforms, and pipes connecting these to shore leak 30,000 tons of hydrocarbons each year. Ships dump 145 million tons of ordinary garbage annually. Salmon, sturgeon, oysters, and haddock have simply vanished. Those fish that survive often suffer from skin infections, deformed skeletons, and tumors.

It is important to educate ourselves about the grave assaults on the biosphere: acid rain, desertification, waste accumulation, overpopulation, ozone depletion. These matter, to know the grim facts about what we humans are doing to nature. But deeper than statistics and reports, it is essential not to lose sight of our love of creation and desire to preserve it. Common sense tells us that if we continue to damage the environment the results will be horrific. Common decency tells us that the greedy, wasteful destruction of life is wrong.

Siegfried Lenz points out that while we appear to live in peace and comfort, at its foundation is a privileged kind of force that is condoned by public authorities and is making our world more and more uninhabitable. Governments and corporations are taking away lakes and oceans, allowing rivers to die, and turning forests to skeletons. A German court declared that those who resist may be acting on legitimate *moral* grounds but are still *legally* in the wrong. As in many countries, in Germany it is business that makes the rules.

Gary Snyder writes that trees and mountains must be represented in Congress and whales must have the right to vote. We in the Greens have tried to further that spirit, to represent those whose voices are not heard in the halls of power—the whales, elephants, dolphins, plants, flowers, and trees of this living planet. The situation is grave. Our hope lies with those friends of the earth all over the world who are rallying to her defense.

—28—
The Politics of Nature

Tom Hayden

The machinery of politics is based on a heartless and hidden debasement of nature. Political scientists speak of the function of politics as the allocation of resources, assuming that nature's bounty is destined to become a Gross National Product for distribution to the powerful. In this way, the dominant politics of growth rests blindly on the death of nature.

Can all this change? Can politics represent ecosystems instead of special interests? Restoration instead of taking? Or does environmentalism have to reject and somehow replace the present political system for its inherent role in causing pollution?

These are questions I ask myself everyday as a state senator. Until recently, not many political thinkers have addressed them. Now, as the environmental crisis begins to overwhelm our past assumptions, it is time for a theory and practice of politics based on restoration and sustainability.

In my lifetime, world population has doubled and natural resources have been cut in half, creating a collision course between industrial growth and the natural world. California is a microcosm, with the fastest-growing population in the U.S. pressing down on a land the first European explorers called "terrestrial paradise."

For example, the development of California has resulted in the loss of all but five percent of the state's original redwood forests. California forestry law promotes "maximum timber productivity" as an official policy goal. That means that a grove of ancient trees, which sheltered our ancestors for thousands of years, which preserved and promoted a living forest for millions of years, is valued only for the price it brings as the deck of a condominium.

U.S. Justice William O. Douglas argued fifty years ago that such trees should have standing in courts of law "to sue for their own preservation." His prophetic opinion has been ignored, and yet corporations, which are utterly artificial entities, are accorded such rights in perpetuity.

This devaluing of ecosystems is reinforced in the case of forests by the vast power of timber interests which, despite their relatively small voter base, can rent the services of willing politicians under a campaign finance system based on unlimited contributions. Corporate contributions are a form of "free expression" astonishingly protected under the First Amendment, at least as presently interpreted by the courts. By contrast, the standing of ecosystems in industrial society is similar to that of the Native Americans or African slaves: a wilderness to be destroyed or domesticated for economic use.

The major strides in environmental law since 1970 must be seen in this context which overwhelmingly favors the utilitarian exploitation of nature. The Endangered Species Act, for example, cannot be triggered until an isolated species is on the edge of extinction. Even then, the dying species must survive for years until bureaucrats and interest groups determine its final fate.

To place a value on nature beyond its immediate use to human society requires a transformation of religious and cultural assumptions that ultimately must be reflected in politics and economics. In this "new paradigm," political economy will no longer be viewed in terms of the machine metaphor of the old industrial order, but in terms of the organic image of a sustainable human community living integrally with nature.

The difficulty of achieving this changed perception cannot be overstated and yet the deepening environmental crisis will compel more and more people to creative adaptation and a rethinking of values. My faith is that the human survival instinct, which forces us to examine our place in the world, is stronger ultimately than the addictive need to cling to obsolete dogmas.

We are entering a transitional era with only two possible endings: either a downward spiral of violent human conflict over diminishing resources, or a spiritual redirection of twenty-first century humanity towards harmony and reciprocity toward each other and the earth. Each of us matters in this unpredictable journey.

Politics is a reflection of pre-existing assumptions about the universe and the role of human beings within it. In technological society, all politics is Newtonian. In his epic *Principia*, the philosopher asserted that "every body continues in its state of rest . . . unless it is compelled to change that state by forces impressed on it . . . and to every action, there is always opposed an equal reaction."

This mechanistic law of the universe was translated into the dialectical materialism of Karl Marx, the atomistic marketplace of Adam Smith, and the theory of checks-and-balances in pluralistic democracies. As James Madison wrote in Federalist Paper 51, "*ambition must be made to counter ambition.*" According to historian Clinton Rossiter, the new American leaders believed that God "had set the grand machine of nature in motion" according to laws "as certain and imperative as those which controlled the movement of the universe." God was, in Rossiter's image, "the great Legislator" of all nature.

In all such theories, political economy was seen as a machinery that rested atop nature, drawing on the environment for raw materials and depositing waste back into that same environment. Such a model externalized nature as somehow "outside" and "below" society, a vast storehouse of resources and debris. In an agrarian society, the model was meant literally too, as in this passage from the Letters of George Logan:

> By political economy is to be understood, that natural order appointed by the Creator of the universe, for the purpose of promoting the happiness of men in united society. This science is *supported by the physical order of cultivation, calculated to render the soil the most productive possible*.

The mechanistic worldview blended with the Genesis mandate into a powerful rationale for subduing the new American continent and its inhabitants; the "savage wilderness" teeming with "savage Indians," as Governor John Winthrop starkly described what lay ahead. The legal argument for taking and exploiting the new lands rested on the Lockean premise that undeveloped land was "waste" or, as Adam Smith described America, "rude" and "barren." Since the Indians had not subdued this *vacuum domicilium* into productive private property, they had forfeited all rights to ownership. The land itself was given value through cultivation by agriculture and mining by industry (for Marx, the "labor theory of value" made the same assumption of nature as only raw material). On a seemingly endless frontier, particularly in comparison with the settlers' densely enclosed European homelands, this vision became our manifest destiny.

The American dream of individual materialist prosperity suppressed a broader dream of the continent. For at least 10,000 years the native people had developed culture, religion, and political economy in reciprocity with their natural world. Unlike the Puritans or Locke, they revered a Great Spirit present throughout the earth community on which their lives depended. Instead of exploitation, they practiced reciprocity towards species around them, from the tiniest flower to the greatest bison. While they sometimes self-destructed or exceeded the carrying capacity of their surroundings, the Native people were on the whole, in the words of a Smithsonian historian, our "first ecologists."

The European conquerors, from Christopher Columbus in the Caribbean to Sir Francis Drake on the California coast, at first observed this intelligent respect for nature in the Natives they encountered. A religious worship of God above and gold below, however, smothered the possibility of an America based on sharing and co-existence between human communities and nature. But if we seek an alternative politics based on sustainability, we must return to our lost continental dream.

The tribes, particularly the Iroquois, believed in a mechanistic governing theory of their own, one of counterbalancing powers, expressed in their structure of confederation. But they also held a view of power as more than the capacity to bend other people or ecosystems to one's will. Power was a creative expression of human energy achieved through an understanding of connectedness with a Great Spirit present in nature. This power was known as *medicine*, a word which itself implies healing. This power was obtained through meditation, ritual, and practice, and was available not only to an individual shaman but to a whole tribe. In many tribes, following the power of a certain chief was a voluntary act, not one of submission and obedience.

Tom Paine encountered this dream around the Iroquois Council fires where the elders discussed their Great Law of Peace. Paine, who was fluent in the Iroquois language, noted in his diaries that "There is not, in that state, any of the spectacles of human misery which want and poverty present to our eyes in all the towns and streets of Europe." Benjamin Franklin and Thomas Jefferson were influenced similarly, the latter describing himself to European courts as "a savage from the mountains of America." Franklin wrote that Indian government was based on a "Council of the Sages, those with powerful medicine," and that there was "no Force; no prisons, no officers to compel Obedience, or inflict Punishment." Jefferson, whose Monticello living room still contains a vast buffalo hide inscribed with a Sioux creation story, pondered the fact that the native Americans never "submitted themselves . . . to any coercive power (or) shadow of government."

These American founders borrowed the Iroquois design of the Albany Plan and the Articles of Confederation. But tragically, they failed to pursue the Indian dream in its community and ecological dimensions. Perhaps it reminded the new Americans too much of their own tribal roots, such as the clans of ancient Ireland, which had been weakened or replaced by the modern nation-state. More likely it was speculative fever that kept the colonists from adopting the Native attitude towards possessions. As Sitting Bull once observed, "the love of possession (was) a disease with them," and thus the early American revolutionaries succumbed to the politics of expanding frontiers. Captive of their times, they perhaps rationalized their choice with the thought that there would always be enough westward American frontier for the preservation of Indian culture and wilderness alike.

The American founders settled on a model of government which was more like the Roman Empire than a tribal confederation. The viability of their project rested on a national psychology of the conquering frontiersman and the availability of unlimited frontiers, symbolized in the nineteenth-century spirit of General George Armstrong Custer. The subject of more paintings than any other general in American history, Custer caused settlers and railroads to invade the great buffalo plains of the Sioux with his fevered reports that the Black Hills were filled with gold. He died "breaking down the gates of Hell" according to one history of the Battle of Little Big Horn. The unrestrained lust for frontier would be a more candid explanation.

In making the fateful choice to be a frontier nation, America's founders consistently rejected the alternative vision of living within limits in sustainable, community-based decentralized governing structures. In the Madisonian/Federalist view, an "extensive Republic" was necessary to provide an outlet for the inherent tendency to factionalism in human nature. Thinkers like Montesquieu, the anonymous anti-Federalist "Brutus" and, in his later life, Jefferson, all believed that democracy could flourish only in a human-scale and sustainable setting. As Robert Louis Stevenson eloquently wrote during the frontier era, "We cannot hope to escape the great law of compensation which exacts some loss for every gain." But the founders' doubts were swept aside in the inevitability of the conquest of the frontier once before them. "Indian society may be best," Jefferson once reflected, "but it is not possible for large numbers of people."

This historic debate remains at the core of political and environmental philosophy three centuries later. The dominant view among Americans and world political leaders is that expanding markets and environmental exploitation are basic to growth and progress, and that environmental mitigation can only be financed from the dividends of such growth. What has changed in our time, however, is the relationship of population to frontier. Where there was a certain logic to boundless expansion in 1775, pursuit of the same logic today is a dangerous and senseless addiction leading ever closer to human ecological destruction.

A new political theory for the ecological era is needed, one which should restore and modernize the older beliefs in decentralization, community-scale institutions, and environmental sustainability which were the dreams of our indigenous ancestors on the continent.

The foundations of an American environmental politics can be recovered from the nineteenth-century Romantics, Ralph Waldo Emerson and particularly Henry David Thoreau, whose visions ultimately animated John Muir towards founding the Sierra Club in 1892. Like the Native Americans, they believed in revelation and transcendence through a spirit within nature. These nature mystics and transcendentalists worshipped the forests and mountains as great cathedrals, not as resources to be felled or mined. They were reacting to the first ravages of commerce created by the railroads and settlers' axes. Some like Thoreau dropped out of Harvard and identified with the victims of expansion, the Native people, Mexicans, and African slaves. All of them wanted to preserve wilderness for its own sake, but also because they recognized in wilderness the essence of the human spirit. In wilderness was the preservation of the world. In place of the machine model of expansionary bureaucracy, they identified with the tradition of the New England town meeting—a model of grassroots, participatory democracy.

It must be noted, however, that these Romantics disdained virtually all forms of politics, a legacy imprinted on later generations of rebels, including environmentalists. Emerson's contribution was to the creation of a national culture. Thoreau adopted self-reliance and self-sufficiency as goals for the individual and community. John Muir came down from the mountains on occasion to lobby elected officials for national wilderness areas; he felt that politics "sapped the foundations of righteousness." Their disdain was reinforced by a context in which politics was dominated by robber barons and ethnic machines at the exclusion of women and blacks who were prohibited from voting.

These environmental founders perpetuated an inadvertent dualism between society and nature contrary to the seamless unity their philosophy embraced. Society meant urban life, a degenerate cauldron of decay, tension, and violence. Cities were best abandoned for the isolation of writing of the solitude of wilderness. Muir condemned the city for engendering a "multitude of wants," and described himself as "withering" every time he visited San Francisco. Our bodies, he felt, "were meant to thrive only in pure air." Muir recommended that people leave the cities "as from a plague."

Thus the Romantics left a mixed legacy for those who seek to ground their politics in environmentalism today. On the one hand, they believed that nature had intrinsic and spiritual value rather than simply a utilitarian benefit. Drawing on Native Americans, they originated an earth-based spirituality in sharp contrast with the narrow interpretation of Genesis employed in defense of aggressive frontier plunder. They also embraced the idea that if politics is to have value, it must be rooted in personal commitment and behavior, arising from below. At the same time, they reinforced a dualistic concept of the environment as "environs," a setting outside the populated areas where social and ethnic politics dominated.

A century later, these ideas continued to have staying power in environmental literature. An eloquent example appeared in a short essay, titled simply "Ecology and Politics," by the eminent environmental philosopher Aldo

Leopold in 1941. As an ecologist, Leopold pointed out that political economy is based on a demand for growth (what he called "take") in excess of
the carrying capacity of nature. The result, he felt, was global war. At the
end of his essay, Leopold asked an intriguing question: *"do other animals select
their 'form of government' to fit their adaptations, or does circumstance dictate the form?"* In
the human case, he seemed to believe, technological society would always
tend toward expansion of form instead of acceptance of ecological limit. In
the spirit of Thoreau, Leopold saw technological society and its politics as
the enemy, ending his essay in an ecological scorn:

> . . . a society rooted in the soil is more stable than one rooted in
> pavements. Stability seems to vary inversely to the mental dis-
> tance from fields and woods. The disruptive movements which
> now threaten the continuity of human culture are born not on the
> land where the take originates, but in the factories and offices
> where it is processed and distributed, and in the capitols where
> the rules of division are written.

There was another stream of environmentalism which has been ignored,
or classified as non-environmental populism, at the turn of the twentieth
century. This was the web of social movements for public health and safer
workplaces, against the squalor, poverty, and deadly pollution of the
swelling cities. Well-documented in Robert Gottlieb's history, these urban
reformers, public health advocates, settlement house workers, and labor
organizers represented a human-centered environmentalism which saw pollution in class terms. Over time, these movements became more social and
political in character, losing their potential linkage with what came to be
known as mainstream environmentalism that focused on the outdoors.

This division effectively deflated the possibility of a successful environmental politics for many decades, cementing in the public mind a separation
between the social and environmental crisis which continues today. The
social movements concentrated on making practical gains within the party
politics and growth based economics of the time, achieving significant gains
for public health and working people within a system that nevertheless continued to degrade nature to create the raw materials of social progress.
Meanwhile, environmentalism was associated increasingly with wilderness
preservation, soil conservation, pollution of resources, and toxic or nuclear
threats to planetary systems. The first, populist environmentalism organized
and reached large voter constituencies while the second, outdoor-based environmentalism raised consciousness but failed to become a powerful interest
group. The two streams connected only occasionally, for example in the late

1950s around the dangerous health effects on humans from atomic testing or pesticide applications. Over the past twenty years these divisions have been healing slowly. Even today, in the semi-factional debates between "mainstream" and "grass roots" environmentalists, or in the scorn of "deep" ecologists towards "shallow" ecologists, there are echoes a century old.

In contemporary times, there are three separate environmental poles of thinking that contend for dominance. First, there are environmentalists working entirely within the present system of politics, trying to create market incentives for environmental reform. These new, respectable pragmatists are gambling on the argument that environmentalism can be proven in business leaders' minds to be more efficient than pollution and waste, and even profitable. Second, there are community- or workplace-based, insurgent environmentalists, often people of color, who are organizing to challenge the narrow market system from the outside. Their assumption is that corporate and bureaucratic power are incompatible with environmental protection for their communities, and must be modified radically. Finally there are more spiritual or New Age environmentalists seeking to re-awaken a fundamental reverence for the earth as a precondition for any change whatever. They seek to educate others to a new vision that is partly mystical and partly scientific, of a living universe and a living earth with a deep spiritual dimension. When this new "paradigm" or "story" reaches enough people, they feel, the structures of society will change towards a new harmony with the earth.

These alternate approaches to the environmental riddle are often successful by their own standards, and they sometimes complement each other in effective ways. But they all are struggling towards a more complete vision and effective practice capable of addressing the urgency of our times.

The problem with the pragmatic view is its faith that leaders of government, business, and organized labor can be "enlightened" by objective presentation to see beyond their short-range interests. The limit of the environmental justice approach is the sheer difficulty of actually transforming the present system of unequal power. The danger in the spiritual approach is that of opting out of external struggles over power altogether, instead concentrating on the inner self alone, tending toward a prophetic sense of irrelevance.

What is problematic politically is that none of these approaches directly seeks to maximize the power of the environmental issue among the majority of voting Americans who consider themselves in some sense environmentalists. To illustrate by contrast, there have been politically powerful single-issue movements on behalf of labor, women, African-Americans, Latinos, Asian-Americans, gays and lesbians—running candidates for offices including the presidency, organizing caucuses to reform party rules, registering voters, and spending money on political campaigns. Environmentalists only lately have grouped together for political action in the League of Conservation Voters and the Sierra Club. Not only have these efforts been recent, but they represent only a tiny fraction of the energy and resources of the national environmental organizations. The NLCV reportedly spent $700,000 in the 1992 elections, for example, while the budgets of the major environmental groups far surpassed $100 million in the same year.

Perhaps there is inherent wisdom in this distance of environmentalism from a politics that does indeed "sap the foundations of righteousness." Where "green politics" has been tried, most notably during the 1980s in Europe, it experienced a brief but spectacular success before floundering on the same divisions I have mentioned. In California, a promising Green Party qualified for the state ballot in 1992, but has imploded in lack of consensus over what to do with itself.

The huge gap between environmentalism and politics, especially in comparison with other social movements, suggests a failure to seize the possibility of shifting the American debate. The fact that Americans in polls consistently indicate strong environmental values is all the more reason to ask whether environmentalism has defaulted on a historic opportunity. Without forming a more viable and organized green voting bloc and running candidates, environmentalists will continue to be marginalized and treated as secondary by politicians of both parties.

What is to be done? I am not suggesting that environmentalists abandon their agendas to become reform Democrats. But they should become a more organized presence, like a "party" but more fluid, in all our stagnating institutions. The difficult task is to achieve broad consensus on an alternative environmental vision and program, followed by a plan of action that not only educates people to the existence of a choice, but threatens and pressures the existing order of power.

Environmentalists must speak of values, alternatives to the present system, and practical steps all in one breath. Values need to be grounded in programs, programs need to be grounded in vision, and both need to be expressed in positive action that arouses people from numbness, denial, and apathy.

The values that we need to articulate are those of Henry David Thoreau against those of George Armstrong Custer. Even in a so-called Information Age, those are the enduring choices. The "information highway" built upon telecommunications technology will not deliver us from the environmental crisis, much less the moral crisis. Ten years after the celebrated birth of the personal computer, which was to "change the world," the global degradation of the environment and human rights continue to worsen, even as they are televised. Now cyberspace is being extolled as "Jeffersonian" by the creator of Lotus, while a "new digital world order" of empowered individuals is advertised by Microsoft.

The more likely truth is that the new information order will be a cyber-extension of the old frontier controlled by multi-national corporations, unless its more revolutionary advocates create democratic "on-ramps" for communities and individuals. And the digital world will be wired by nuclear power and coal plants unless solar energy is brought to earth instead of powering space satellites where utility lines cannot reach. To define environmentalism only as the new efficiency, as enlightened corporatism, will accelerate recycling and green consumer products, but completely misses the point that our very survival is threatened unless we undergo a serious process of conversion to saner values for the ecological age.

But we cannot follow Thoreau's path into withdrawal. The concepts of community and personal self-reliance, environmental restoration and sustainability, reverence for nature instead of things, must be translated into practical efforts to reform and eventually transform a completely dysfunctional system. Here the work is endless, and the examples many, including:

* Revision of school and university curriculums to integrate ecological values and science throughout the process, as exemplified by the work of Lynn Margulis and Carolyn Merchant (Why don't environmentalists run for school boards instead of yielding to the fundamentalists who still fight Darwin?).
* Reform of economic models to internalize environmental costs and replace GNP-measurements with quality-of-life ones (Herman Daly, Hazel Henderson, Paul Hawken).
* Restructuring of economic incentives and regulations to penalize pollution, waste, and throw-away consumption patterns; mandatory recycling and reuse of all wastes; reliance on renewable resources in place of non-renewable fossil fuels; and the creation of products which last. Such changes would be possible only with a transition to taxes on waste, consumption, and pollution instead of income and payrolls (Al Gore's book, and Hawken again).
* Extending the democratic process to the marketplace and corporate bureaucracy by empowering individuals to participate in decisions on the basis of greater access to information and technology (the enduring vision of Ralph Nader).
* Restoration of democratic government by campaign finance and lobbying reform to weaken the stranglehold of oil, chemical, agriculture, and other

special interest groups over the fate of environment and society (Jerry Brown).

* Redesign of representative electoral districts to include natural bio-regions as well as voter blocs, and restoration of habitat for such species as salmon and grizzly (the politics of Gary Snyder, the writings of Schumacher).

* Acceptance of "the rights of nature" (the concept of Justice Douglas, the phrase of Roderick Nash) as the next great extension of the moral community of respect that democracy represents.

* The return of the spiritual to politics as in the tradition of the Great Council Fire, the end of Government as Leviathan (Thomas Berry, Joanna Macy, Matthew Fox, and Al Gore).

No one of these efforts at transformation is more important or more urgent than the others. But none of them alone is sufficient for the great transition, indeed the evolutionary leap, that we need to accomplish. Taken together, they are *strands of a single process that unravels and reweaves our whole social order at the same time.*

This agenda is an alternative to both withdrawal from political engagement and piecemeal efforts at reforming a hopeless system. The alternative for the environmental movement is to develop a "medicine" stronger than the coercive power and escapist fantasies of the present growth machine.

It means working to transform ourselves from frontier personalities to citizens at home in our communities and bio-regions. It means transforming ourselves into green consumers and voters. It means transforming the institutions where we work and worship into centers of ecological action. It means storming politics—and all the castles of the obsolete—with a transforming spirit, a *ruah*, breath of God, that brings an earth politics to life. Such an earth politics, a *down*-to-earth politics, cannot be co-opted or denied. Such a politics overcomes the love of power with the power of love, the logic of power with the power of logic—but also with millions of people ready to vote, as Thoreau did, "not with a mere piece of paper, but your whole life."

—29—
The Age of Light

Beyond the Information Age

Hazel Henderson

The "Information Age" has become a ubiquitous image among futurists seeking fruitful metaphors for the ongoing restructuring of industrial societies. New images of post-industrial society proliferated in the 1970s as the restructuring of industrial societies accelerated. Alvin Toffler's *The Third Wave* (1980) depicted a globalizing human culture based on information, more appropriate use of technology, and more productive, pro-active individuals outgrowing consumerism. In *Alternative Futures for America* (1970) and several other books, Robert Theobald[1] outlined what he called "the Communications Era," describing a shift toward community empowerment, automation, and great dissemination of information.

My own view is that information disseminated more broadly could facilitate the rise of networks of citizens about to cross-cut old power structures, facilitate learning, and initiate a widespread politics of reconceptualization, transforming our fragmented world view into a new paradigm based on planetary awareness: an Age of Light, auguring planetary cultures sustainably based on renewable resources, and a deeper understanding of nature. At a more basic level, we know from quantum physics and Planck's constant that it is *light* (rather than matter, energy, or time) that is fundamental to the universe. Planck's equation holds that quanta of light are also quanta of *action*. The Age of Light has been augured in most religious literature, most notably in the Bible. In Genesis, God's first command was, "Let there be Light."

The Age of Light lies beyond the Information Age. The Information Age is no longer an adequate image for the present, let alone a guide to the future. It still focuses on hardware technologies, mass production, narrow economic models of efficiency and competition, and is more an extension of industrial ideas and methods than a new stage in human development. Information is an abundant resource rather than a scarce commodity (as in economic theory) and demands new cooperative rules from local to global levels.

Leading Edge Technologies Mimicking Nature

Some examples of existing successful technologies based on nature's design: *cameras* mimic the eye; *airplanes* mimic birds; *radar* and *antennas* mimic insects

Information Technologies:

- ARTIFICIAL INTELLIGENCE . . . Expert systems, "hypertext," associative learning program
- BIOTECHNOLOGIES . . . Genetic engineering, cloning
  - . . . Monoclonal antibodies, interferon, insulin
  - . . . Gene splicing, hybridization, tissue culture
  - . . . Luciferase
  - . . . Pheromones, chemical attractants, biological pest control
  - . . . Protein-based catalysts, assemblers, microbes that "eat" oil spills, sulphur, etc.
- MOVING BEYOND MEDICINE THAT AUGMENTS THE BODY'S DEFENSES AND HEALING PROCESS TO CELL REPAIR AND LIFE-EXTENSION
- ENERGY TECHNOLOGIES . . . Ocean thermal, tidal and wave generators
  - . . . Biomass energy conversion
  - . . . Dams, hydropower
  - . . . "Nanotechnology," molecular assemblers
  - . . . Synthetic photosynthesis (zinc pophyrin and ruthenium oxide), photovoltaic cells
  - . . . Osmosis, membrane technologies
  - . . . Solar arrays and sails
  - . . . Fusion reactors

Nature's Models:

- HUMAN INTELLIGENCE, KNOWLEDGE
  - . . . Human memory, language
- DNA, RNA CODES, VIRUSES, BACTERIA
  - . . . Human immune system
  - . . . Plants, wild species
  - . . . Fireflies
  - . . . Insects, microbes, fungi
  - . . . Amino acids
  - . . . Microbes
- HUMAN IMMUNE SYSTEM, DNA, GENES

- OCEANS AND OTHER GLOBAL PROCESSES
  - . . . Natural decay processes, fermentation
  - . . . Gravity
  - . . . Viruses
  - . . . Green plants chloroplasts
  - . . . Living cell membranes
  - . . . Insect wings
  - . . . The sun

PLATE I

Interconnectedness Increasing

Information itself does not enlighten. We cannot clarify what is *mis*-information, *dis*-information, or propaganda in this media dominated, "spin-doctored" environment. Focusing on mere information has led to an overload of ever-less meaningful billions of bits of fragmented raw data and sound bites rather than the search for meaningful new patterns of knowledge.

My view of the dawning Age of Light involves a repatterning of the exploding Information Age. This requires nothing less than a paradigm shift to a holistic view of the entire human family, now inextricably linked by our globe-girdling technologies. The earth is re-perceived as a living planet, and the most appropriate view is organic, based on the self-organizing models of the life sciences. Biological sciences become more useful spectacles, and it is no accident that biotechnologies are becoming our most morally ambiguous tools.

The biotechnology revolution has been in high gear since the late 1970s, when entrepreneurial biology professors began to make commercial joint ventures in genetic research and engineering, gene splicing, etc., which led to the boom on Wall Street in stock of such companies as Genentech, Inc. Ironically, most of the research that led to the biotech industry had been paid for by U.S. taxpayers—who were elbowed out of sharing in the huge profits from many of these deals. The most effective crusader for the public interest in this field, its ethics and hazards, has been Jeremy Rifkin, who sounded the alarm with his books *Algeny* and *Who Should Play God?*, co-authored with Ted Howard.[2] Rifkin challenged untried genetic experiments in court and fought a valiant court case to prevent corporations from "patenting" lifeforms—a battle lost in 1987, when the U.S. Supreme Court extended the original U.S. Patent Office ruling allowing limited patenting to all life forms. Thus as Rifkin points out in his *Biosphere Politics* (1991) the whole planet's genetic heritage, including DNA itself, which humans share with other species, can now be turned into private property.[3] Human genetic material can also be manipulated, combined with that of pigs, sheep, and mice, as is occurring today. One inspired protest was dreamed up by Gar Smith who in 1980 began circulating mock U.S. Patent Office applications to human parents, so that they could "patent" their own offspring—to spoof the absurd Supreme Court ruling.[4] Now the polymerase chain reaction (PCR), a method of making millions of copies of a piece of DNA quickly and reliably, has been sold by Cetus Corporation of California to Hoffman LaRoche, the Swiss multinational pharmaceutical firm. "PCR opens up whole new areas and techniques." As reported by *The Economist*, "It is the key to the biological candy store."[5]

The Age of Light will follow on from the Solar Age as humans gradually learn that it is light and action that are fundamental in the universe. The technologies of the Age of Light are already appearing. Beyond electronics, these phototronic technologies are miracles of speed and miniaturization, such as the new 0.25 micron "superchip" (four hundred times thinner than a human hair). These superchips now on the drawing board can pack hundreds of millions of transistors—ten times today's record. In the year 2000, 0.1 micron widths will be the cutting edge, small enough to cram billions of transistors on a single chip. Even today's advance optical printing methods will have to give way to high frequencies in the spectrum, using x-ray lithography. In 1991, San Jose, California's Cypress Semiconductor Corporation announced (*Business Week*, June 3, 1991) that it had purchased from Hampshire Instruments, Inc. of Rochester, New York, a new system that generates the x-rays, not with a room-size synchrotron (or atom smasher) that costs $30 million, but with a laser costing a modest $4 million.[6]

Another aspect of these new technologies of the Age of Light is the level of integration achieved with the biological sciences, as they shift from the "inert" classical physics world view to the organic living system perspectives of biology and ecology. From Plate I ("Leading Edge Technologies Mimic Nature") we see striking confirmation of the continued inspiration to nature in their design. However far out and "high tech" they all claim to be, and indeed are, they, like most technologies before them, owe their inspiration to nature. After all, Gaia, the great designer, has been optimizing these energy capture-utilization-storage systems for billions of years. Gaia is also the pre-eminent innovator and experimenter, who excels in sheer artistry as well. Further evidence that humans are still imitating nature is the growing industrial research field of biomimetics (literally "mimicking nature") described in the *Business Week* cover story, "The New Alchemy."[7] For example, new ceramic materials mimic the molecular structure of abalone shells to make impact resistant armor for military tanks and aircraft. As the article points out, "Nature's materials have passed the test of time." K. Eric Drexler's vision of "nanotechnology" in his 1986 book, *Engines of Creation*,[8] predicted the shape of much of his new biomimetic research, such as Argonne National Laboratories work on improving structural materials by "flirting with Creation," developing entirely new materials atom by atom using "nanophase materials" such as crystals less than 100 nanometers across—smaller than most viruses. Japan's Ministry of International Trade and Industry (MITI) has launched a ten-year program on "nanotech initiatives" which it hopes industry will fund to the tune of $100 million or more.[9]

The Age of Light is an image that reminds us that it is the light from the sun that drives the earth's processes and powers its cycles of carbon, nitrogen, hydrogen, and water and the climate "machine." It is these light-driven processes—which are then mediated by the photosynthesis of plants—that maintain conditions for us to continue our evolution beyond the Information Age. Our present technologies are already maturing from their basis in electronics and are shifting to phototronics. These new light-wave technologies include fiber optics, lasers, optical scanning, optical computing, photovoltaics, and other photoconversion processes.

As we progress in these areas we will notice how each one leads us into a deeper appreciation of nature's technological genius; we have modeled our earlier breakthroughs, such as flight, on that of birds. Nature's light-conversion technologies, the basic of which is photosynthesis, still serve as design criteria and marvels of miniaturization, such as the chloroplast cells all green plants use to convert photons into usable glycogen, hydrocarbons, and cellulose. This is still the basic production process on which all humans rely and when our photovoltaic cells can match the performance of the chloroplast, we will be on the right track. Thus the Age of Light is more than the new lightwave technologies emerging from the computer, robotics, and artificial intelligence labs. The Age of Light will be characterized by our growing abilities to cooperate with and learn from nature. The Age of Light will build on today's biotechnology, still in its exploratory, often exploitative, moral infancy.

The Age of Light will bring a new awareness and reverence for living systems and the exquisite information technology of DNA, the wisdom and coding of all living experience on this planet. The Age of Light will go far beyond industrial, manipulative modes toward deeper interconnecting, co-creative designing with and learning from nature, as we become a species consciously co-evolving with all life forms on this unique water planet. The Age of Light will also be one of a time compression as we include our holistic, intuitive, right-brain hemisphere cognition with our more analytical, left-brain functioning, and as our computers catch up in their abilities for parallel processing with the simultaneity of our own brains' synapses. The peerless design of the human brain presents the ultimate challenge to computer designers, despite the much-vaunted progress in so-called "artificial intelligence" systems. For example, a leading designer of Cray Computer Company's efforts to create a supercomputer mimicking more closely the parallel processing ability of human brains, recently left to pursue other areas of endeavor. Most leading edge technologies based on light, whether information, solar energy, or biotechnologies and gene splicing, also mimic nature's design, such as monoclonal antibodies, interferon, and other methods based on learning from the human immune system. The "nanotechnologies" or protein-based assemblers that Eric Drexler (mentioned earlier) envisions, are designed by mimicking the functions of amino acids.

At the same time we are learning much more about our own bodies' responses to light, and how humans deprived of full-spectrum, natural light in indoor living and working conditions can suffer weakening of their endocrine and immune systems. Thus the wisest of us recognize that our earth still has much to teach us—if we can humble ourselves and quiet our egos long enough to really listen and see, hear, smell, and feel all her wonders. When we can feel this kind of attunement to the whole creation, we are transported with natural delight to the "high" that psychologist Abraham Maslow called "peak experiences." As we reintegrate our awareness in this way, we no longer crave endless consumption of goods beyond those needed for a healthy life, but seek new challenges in our societies for order, peace, and justice, and to develop our spirituality. It is in this way that humans can overcome the dismal Second Law of Thermodynamics in the continual striving for learning and wisdom. We no longer blind our imagination with the dismal deterministic view of a universe winding down like a closed system. Since Prigogine, we know that the universe is full of surprises, innovation, and evolutionary potential.[10] In fact, Cartesian science's search for certainty, equilibrium, predictability, and control is a good definition of death. We should happily embrace the new view that uncertainty is fundamental, since it also implies that everything can change—for the better—in a twinkling of an eye! As we move on to post-Cartesian science, we can acknowledge the earlier period of the scientific enlightenment of Descartes and Newton, Leibnitz and Galileo. Its instrumental rationality and manipulation of nature did lead to that greatest outpouring of technological hardware and managerial virtuosity which we call the Industrial Revolution.

As we have seen, the whole process of human development is teleological and evolutionary, and therefore cannot be explained or predicted by existing reductionistic scientific paradigms. This great purposeful unfolding of human potentialities toward goals—bettering human societies, perfecting the means of production, and fostering conditions of people's lives so that they might fulfill themselves—is essentially a spiritual, as well as an instrumental and materialistic endeavor.

Binding such a transcendent set of human goals and visions of the future within the so-called "laws of economics" was and is a travesty. With new perspectives and new paradigms in the 1990s we can move beyond old conceptual prisons, whether the reductionist view of the Information Age or the, so far, literal interpretations of the Solar Age. When *The Politics of the Solar Age* was first published in 1981, it was catalogued among energy books. Many who heard my lectures asked me such questions as, "What percentage of my house's heating needs could be met by solar energy?" I would always reply that this was only one level of the meaning of the Solar Age, and then proceed to remind my questioner that if the sun was not already preheating their house and the earth, it would be about 400 degrees below zero! Thus I was referring to many other levels of the transition, from the Age of Fossil Fuels to renewable forms of energy, to the needed design revolution in our technologies and social structures, as well as to all the cultural and mythic levels of such a planetary transformation. Since then, I have focused on the evolutionary process of human development as the frontier of all these needed changes.

Human development and social organization are processes that, by definition, have goals, purposes, and values—and move toward them. Thus any discipline still based on classical physics and Newton's celestial mechanics, particularly economics, cannot be overhauled or expanded enough to map the dynamics of such unfolding processes. In fact, Western science has many specific *prohibitions* against any hypothesizing about values and purpose, and most often denies the existence of teleological aspects in nature. Thus, its methods are based on the search for certainty, fundamental laws, and exactitude (what I have termed "micro-rationality" to distinguish these endeavors from the mapping of larger contexts and more holistic inquiries, which I term "macro-rationality"). A hopeful sign of today's shift to broader paradigms is the fact that the process of human development is being referred to less and less often as "economic development." The word "economic" has been dropped altogether in the new ecological definition: "sustainable development."

Meanwhile, classical science still has no theory of *process*, and development is a multi-dimensional process. Arthur M. Young in *The Reflexive Universe*[11] presents a sweeping synthesis of science and human development in his Theory of Process, based on quantum, as well as classical physics. Young, the developer of the Bell helicopter, assumes that the universe is based on freedom and that the fundamental laws that humans have discovered are constraints on this freedom, but they are secondary. As any artist or designer knows, constraints actually serve the creative process—providing the medium, conditions, and context of the play of creativity. Young believes that the problem with the main body of Western science is that it limits itself to focus on the material plane, dealing with "inert" matter, i.e., molar objects, such as rock. (These appear "inert" because the random motions of their molecules cancel each other out.) Young's Process Theory starts not from matter as fundamental, but with light, or action as fundamental (i.e., Planck's constant, as mentioned, views light as quanta of action, also quanta of uncertainty). From this beginning point, Young assumes that processes are, by definition, purposeful and involve *individual* actions of atoms, molecules, and organisms with inherent goals (rather than the statistical probabilities or averages of the classical view).

Thus, the Age of Light is metaphoric to many levels:

First, it is based on the re-membering and re-wholing of human perceptions and paradigms. Nowhere is this now more evident than in the wide understanding of the first law of ecology: Everything is connected to everything else. I discussed this and its complementarity with the new view in physics, citing Bell's Theorem implying a non-local universe (i.e., action and interactions occurring at a distance, with no apparent medium or linkages) in the last chapter of *The Politics of the Solar Age*. At the same time, I proposed a post-Cartesian scientific world view where changes and uncertainty are the new constants, together with principles of redistribution and recycling of all elements, heterarchy, complementarity, interconnectedness, and indeterminacy.

Secondly, this new paradigm and world view also represents a new synthesis between Western science and religious and spiritual traditions, since it embraces purpose and meaning as fundamental to life processes. Furthermore, purpose implies cognition, consciousness, heuristics, goal-seeking, and visioning as central to all life, at various levels, and most pronounced in humans.

In previous ages of human history, the early gatherers and hunters learned to be competent in the material world by using and storing naturally available photosynthesized plant energy stored in berries, seeds, and trees. These early humans also invented the storage technologies—pots and woven baskets. They worshipped the Great Mother, symbol of the earth, and as I have proposed, also symbolic of the species gene pool. They lived in matrifocal communities and were generally (as pointed out by Riane Eisler in *The Chalice and the Blade*[12]) peace-loving, with males and females in equal partnership. I have also termed this early Neolithic period the Age of the Genotype, since the species' needs came before any individual phenotype—indeed, humans at this point may not have been significantly individuated or aware of themselves.

In the Agriculture Age, humans began to learn the many ways of augmenting naturally occurring energy and accumulated it, intensifying natural photosynthesis by farming and domesticating animals. This period ushered in the Age of the Phenotype, where the rule of the Great Mother and the values of the gene pool were eclipsed by the newly individuated phenotypes seeking their own meaning and experimenting with the natural world. This process continued right through to the Industrial Age, where humans mined out much of the planet's sixty-million-year-old endowment of fossilized solar energy—oil, coal, and gas (laid down by plants) in an unprecedentedly brief period of less than three hundred years. In spite of all the mistakes, we can only honor the impulse of individual humans to act in freedom, to test out their physical and mental skills and their increasing power to manipulate nature. British science writer Nigel Calder portrays the positive aspects of the biotech revolution in *The Green Machines*.[13]

Thirdly, the dawning of the Age of Light augurs the current reintegration of ourselves into a new level of awareness of the needs of the gene pool, as we engender ever greater risks to future generations with our technologies. Thus there is also a reawakening of the values of the Great Mother and a concern to rebalance gender roles and responsibilities in a new partnership society, with cooperation and peaceful conflict resolution now clearly the *sine qua non* of our survival. Gregory Bateson's *Mind and Nature* provides us a valuable synthesis.[14] Naturally, such a vast cultural change has also ushered in a new search for more comprehensive meaning to decode all of our accumulated cultural DNA in finding a new place for humans in the cosmos and new significance for the human journey on this earth.

The Solar Age and its politics have also arrived—a decade later than I had predicted—given ghastly impetus by the 1991 Gulf War. This tragic, unnecessary conflict, which could have been avoided, will be viewed as the inflection point of fossil fueled industrialism. The politics of the shift to the Solar Age is fundamentally different from that of the industrial period. The insights of English chemist Frederick Soddy (1877-1956) who shared the Nobel Prize with Rutherford (*Cartesian Economics*, 1922) could have corrected economic theory for the Solar Age,[15] but he was ridiculed by the economists of the time and had to self-publish his book. The Solar Age will be more decentralized and the transition to more democratic processes will be necessary. Hierarchical mega-governments and mega-corporations, continent-wide transport and distribution of food and goods were all predicated on cheap oil. The underlying dynamic of the Solar Age, as Soddy predicted, will be how to control energy flows in human societies most efficiently and implies a top to bottom design revolution in all societies, which is at last beginning.

The forces and lobbies of the past—nuclear and fossil energy companies, interstate highway builders, concrete pourers, automakers, and all the other industrial sectors built on waste, maximum energy use, and rapidly increasing levels of entropy—are now, all over the world, locked in legislative and market combat with the emerging sectors. The new industries which minimize entropy are based on refined, miniaturized, and more "intelligent" technologies which pinpoint end-use energy needs, re-use and recycle all resources, and tackle the job of cleaning up the devastating effects of industrialism. The newest enterprises must address environmental restoration and enhancing where possible the performance of ecosystems. In 1991, one of the first "trade shows" of this budding sector of twenty-first-century economies was convened in Florida by the Society for Ecological Restoration. Here biologists replanting sea grasses rubbed shoulders with ecologists from power companies in charge of remediation of spoiled lands and specialists in restoring ruined soils with specially designed crops or micro-organisms. Bio-remediation and desert-greening will be big business in the twenty-first century. Of course, this whole scenario will depend on capturing more elegantly and efficiently some of the planet's abundant daily photon shower from the sun. As Soddy put it in 1992, "how does man, or anything live—*by sunshine!*"

This will require that economics be demoted from a macro-policy tool to a micro role in keeping books between firms, based on full-cost pricing. The *data* on externalities and social costs would have to be developed by more realistic disciplines: thermodynamics, biology, systems and chaos models, and ecology. For example, as economic growth models disordered and destroyed ever larger ecological and biospherical systems, the analysis of this damage was taken over by such interdisciplinary teams of scientists as the U.S. National Committee on Man and the Biosphere (MAB) and its program for evaluation of Human Dominated Systems (i.e., those significantly affected by human activities). MAB's programs, started in 1989, include in its 1991 agenda a U.S. Action Plan for Biosphere Reserves. Economists are *not* included in most of its scientific committees, since its central concerns are of ecological sustainability.[16] As I have emphasized for the past twenty years, *these* are the criteria that must form the context for all human activities and "development" goals—rather than *any* of the unscientific formulas of economical theory, still based on the outmoded "welfare" theory, Pareto Optimality (which ignores income distribution and asks absurd questions, mentioned earlier, about how much people are "willing to pay" to preserve a wilderness area from developers who can always outbid citizens, because the "development" will reward them rich profits). Allen Kneese, always one of the rare breed of honest economists (in the Soddy, Georgescu-Roegen, and Daly tradition) reviews the limitations of "environmental economics" in "The Economics of Natural Resources."[17]

The dire consequences when ignorant humans trespass into "managing" natural systems is now clear, for example, adding some seven billion tons of carbon to the earth's atmosphere each year—contributing to global warming—while at the same time cutting down between thirty to fifty million acres of carbon-absorbing forests. World Resources Institute of Washington, D.C., offers some strategies to check this looming disaster. Tree planting is a primary strategy—one adopted all over the world by ordinary citizens. While serving as a member of the Advisory Council of the U.S. Congress Office of Technology Assessment in 1975, scientific committees were still arguing about whether there was a problem. In 1989, the experts and politicians caught up with citizens' movements such as Chipko and Africa's Greenbelt tree-planters, when environmental ministers from sixty-eight countries signed the Noordwijk Declaration (in the largely below-sea level Netherlands) which called for the goal of an annual thirty-million-acre net increase in forest cover by the year 2000, to reduce net human carbon emissions.[18] Thus economists are learning that investments in the 1990s will have to be directed into ecological restoration and social programs—health care, education, and population control—a far cry from their current investment priorities to hype GNP growth.

The Age of Light will also be a time of re-weaving the world's cultures, most of which down through history have understood our planet's total dependence on the sun. They understood and worshipped our Mother Star in myths and traditions, as well as the primal light of the sun, whether from stars or distant galaxies. As physicist David Peat notes in *The Philosophers' Stone*, synchronicity is the bridge between mind and matter, and he lovingly describes the new sciences of "chaos" now inching toward mapping some parts of this wonderful universe where light, action, and surprise rule, all within the constraints so well mapped by classical physics.[19] Peat envisions an era of "gentle action," where millions of more aware people in more democratic societies can act more intelligently together. As Arthur Young adds, "science is humanity's map, but myth is its compass."

The Age of Enlightenment, some three hundred years ago in Europe, expressed some of these hopes and visionary designs for human potential and development. As we move beyond the Information Age to greater wisdom, we may steer through today's crises and clouds into the sunshine of the Age of Light.

—30—
On Sustainability

B. D. Sharma

(The author, till very recently the Commissioner, Scheduled Castes and Scheduled Tribes, Government of India, was chosen, on retirement, to be the Chairman of the Bharat Jan Andolan or People's Movement of India, a conglomeration of scores of people's movements.)

Dhorkatta, Bastar, Madhya Pradesh, India—May 1, 1992. Honorable Members of the Earth Summit, Rio, Brazil: We, the residents of this small village republic, deep in the luxuriant subtropical forests of the Indian sub-continent, wish to invite the attention of your august assembly to some vital issues concerning "the future viability and integrity of the Earth as a hospitable home of human and other forms of life," the main theme of your deliberations at Rio. Before we begin, however, we profoundly compliment you on your bold initiative in holding the Earth Summit. At this end of the globe, in our small forest habitats, we too share your fears for Spaceship Earth.

You should know that in our villages we have stopped, totally, the commercial exploitation of our forests. The government of our country, of course, may not appreciate the spirit behind our decision. They have, in fact, taken it to be defiance of the law. For the forests formally belong to the state. We are, accordingly, treated as intruders in our own abodes where we have been living through the ages. Consequently, according to the law, we cannot even dig for roots and tubers, pluck fruits, or even breathe freely the nectar of earth. We cannot pick bamboo to cover our huts, or cut a pole to mend our plough. "That will destroy the forests," they say. And when magnificent tall trees of all varieties are mercilessly felled and carted away, leaving the earth naked and bare, we are told that is scientific management. That such acts are performed in the service of the nation. The little sparrow and owl meanwhile desperately flutter about searching for a place to perch. But even the hollow trunks of dried trees have not been spared!

This perception of national economy which the state today represents is not the perception of the people for whom forest, land, and water together comprise a primary life-support system. The legal fiction of the state's suzerainty over natural resources was created during the colonial era and has been continued and even reinforced after independence, in the name of development. This is not acceptable to us. It is a denial of the very right to life with dignity—the essence of a free democratic society. We are confident that this perception of ours is shared by the people similarly placed across the globe.

We, therefore, respectfully submit that the honorable representatives of governments at the Earth Summit are not competent to speak for the disinherited among us. Your perceptions and therefore your stand will be that of estate managers keen to exploit resources on the lines already set by the North. In the past this has invariably implied deprivation of the masses to benefit small elite groups. Frankly, we fear that even though the honorable representatives of Non-Government Organizations (NGOs)—notable exceptions apart—may differ in their views with the state, they are, by and large, bound to share such common basic frameworks as are necessary for their acceptance as partners in the negotiating process. It will not surprise us, therefore, if deliberations at the Summit turn out to be partial. In which case the conclusions will almost certainly be one-sided. This fear is amply borne out in the way the agenda has been framed and also by the Prepcon discussions.

The rich countries are justifiably keen that natural tropical forests be preserved. We too feel the same way, but for different reasons. You require "sinks" for the carbon dioxide emitted by your automobiles which are vital for your "civilization on wheels." We hear about a queer proposal for the declaration of our forests as "global commons." Forests as wilderness would be ideal for this purpose, though you would not mind enjoying usufructory rights. But our paddy fields will be out of place, for they produce CO_2 and thus compete with your cars for that sink. So your basic position as far as we can see is identical to that of our governments. In both cases the people themselves are dispensable. In truth, the two are virtually one as the modern sector of our country, for all practical purposes, is a mere extension of the Western economic system. Of late, in fact, even the thin veneer of national identity has been blown away by the gusty winds of globalization. Discussions at the Summit are bound to be in the nature of bouts for booty rather than for responsible handling of a sacred trust of humankind—generation after generation. But this can be avoided. Please give what follows a patient and considered hearing.

Friends, we are surprised at the casual and parochial vein in which grave issues concerning the survival of life itself have been taken up. If you fail, nothing will remain. If nothing remains, what will be there to share and fight about? But this is the way of all estate managers. They must assume they are always right. Our own experience, a very bitter one, bears this out. In the name of preservation of forests, for instance, our ancestors were mercilessly driven out. And what followed in the name of scientific management was catastrophic. Luxuriant natural forests which sustained us were replaced by teak, which does not even provide us shade in summer. Then came eucalyptus under which not even grass can grow! After that, it was the turn of vast plantations of pine, which would burn like a torch in high summer. But each of these decisions was proclaimed as *the* right way. And to question such projects was blasphemous. Tragically for us, the estate-managers never recognized that the true worth of the magnificent sal, *Shorea robusta*, was far, far greater than the cash recovered from a dead log. The sal is *Kalpavriksha*, the tree that fulfills all desires. Once sal vanished from our forests the struggle of forest dwellers became reduced to physical survival—the evening meal. You see the irony. You worry about how your cars and air-conditioners can continue to operate for a hundred, or a thousand years. Our concern is the next meal that has to be procured at any cost. How can these two perceptions ever meet unless *you* see things from our end of the world and set your own perspectives in order. Friends, can you really not see how far such trivial priorities as air-conditioning and aerosol, with all they represent, have pushed the earth? Yet, you continue to talk about business as usual, of development through your lens, fueled by the same ecological system which has pushed us to the brink of an ecological abyss. Worse, you pose poverty as the worst pollutant and dedicate most of your agenda to eradicating this "environmental hazard."

On the face of it your endeavor might well sound laudable, but consider the hackneyed prescriptions you have chosen to tackle poverty—management of capital, technology, and resource flows. There are two reasons why this framework does not sit well with us. Nor, incidentally, can it help you in the long run. Let's first take the economic frame. Be clear that the phenomenon you are talking about has little to do with poverty. The issue is one of deprivation and denial. You seem ignorant of the fact that we have been robbed of not only our resources, but the great wealth of our life-sustaining skills acquired over millennia. Seen from your horizon, ordinary people are ignorant. Even despised.

Why are we despised? Because we live closer to nature, we do not don many clothes, nor do we have much use for your kind of energy options. We are, therefore, "'poor" in your book. And since you, with missionary zeal, wish to "eradicate poverty" we must be enabled to acquire more commodities, consume more. Is this not why the czars of your ecological system incessantly bombard us with visuals of the glittering life? Making our simple ways look ridiculous by contrast to your own may well whip up new demands and expand your markets, but can you seriously suggest this to be the way to eradicate poverty? Such approaches have been directly responsible for the phenomenal inequality we see around us today. These are also the very reasons that the ecology of vast portions of the globe has been so terribly fractured. Yet, the estate managers of the world continue to wrangle for inflated entitlements and deflated obligations, indulging in reckless brinkmanship in dealing with the commons.

This, friends, is the law of the market. Little wonder that the focus of the Earth Summit has already shifted from land, water, and air, to the illogical issue of money! This drift, to our minds, is contemptible.

Those who have crossed the Rubicon of consumerism must point out at this stage that the Summit debate seems poised to miss the main point. You are no doubt talking about the quality of life, but *within* the consumerist paradigm of development and bounded inescapably by an economic framework. Other aspects of life have not even been brought up. We do not blame you for this lapse, for as leaders of the "modern" economic world, you have no experience of the "real life." In a bid to make the system produce more, for that is what decides its competitiveness in the market and its ranking in the world, all that is human is squeezed out, bit by tiny bit. Human concerns and relationships are dispensable, or at best market-convertible. Rushing to the faraway home to be by the side of an ailing mother, leaving the working machine unattended, is not rational. "Do not get emotional, you are not a doctor, send money instead," counsels the manager, worried by the high incidence of absenteeism in his production unit. To us this is the advice of an eccentric. To you it is the cold logic of your economic system. The machine *must* be used round the clock or else you lose your competitive edge. And people? They are but extensions of the machine! For them, even sleeping at night represents lost opportunities! But you have designed ways for the rattled living robot to enjoy "perfect equanimity." A variety of vintage spirits, or still more modern aids such as heroin, cocaine, and LSD are on hand. At the end of the day, the market determines the cost of life and living.

Look again at your world. The community has already been sacrificed on the altar of productivity. The family now is the last impediment in the way of achieving "perfect rationality" and highest levels of productivity. But even here solutions were at hand. Within the family you dispensed with the burden of dead wood by packing your elders, where necessary, to senior citizen's homes. Now only the nucleus of husband-wife remains, at best. But this too appears to be haunted. Why should a man and a woman remain tied by emotional bonds for life? They too must subjugate themselves to the dictates of the economic system, each one serving the system at points most suited to it. Thus, marriage must break. Living together is good enough for sex. And sex, of course, can be rationally negotiated in a free market. The recent trend towards cynicism about motherhood and about women having eternally to carry the cross of procreation is really the culmination of the challenge of reason against human emotion. Such are the compulsions of perfectly rational beings.

Can you recognize the ugly, twisted logic of your economic system? Perhaps it is too much to ask. For you are clearly dazzled by its benign aura. You have surrounded yourselves and studded your abodes with all sorts of gadgets—surrogates for human concerns, relationships, and emotions. Even your moments of leisure, acquired at heavy financial costs, are determined once again by the market. You are no longer able even to laugh and dream unaided! Having lobotomized the soul from your neighborhoods you now take refuge in the mirage of telecommunications and rapid transport to create the illusion of "one earth."

But let us, for the moment, set aside human concerns and relations. Instead, let us consider the implications of this market-substitution which the economic system is coercing the rest of us to emulate as a lifestyle model. Given the proclamations of your scientists and even some of your world leaders, you obviously admit that we are poised on the brink—even before one in five people (who command four-fifths of the earth's resources) have been able to attain the desired standard of life. How much further must we continue to tread the same lethal path before the final collapse? This is the question the Earth Summit must ponder. Can you really not see the catastrophe you have set into motion? Having "co-opted" your own elite, you state that poverty alleviation is now your objective. This is the mirage we are condemned to chasing in vain, endlessly. Meanwhile you content yourself in tinkering with buttons, watching us follow in your footsteps even as a void engulfs us and our communities and families shatter. The writing is on the wall. The omnipotent, omnipresent market is turning living, breathing men and women into commodities-in-trade.

Honorable members of the Summit: it is in the face of this deluge that we earnestly call upon you to put you agenda, indeed your houses, in order. The development and associated lifestyles you chase are a hallucination. There is nothing sustainable about your ambitions. Your blueprint of sustainability will not even nourish a tiny section of humankind. Ironically, even as the bulk of humanity suffers hitherto unthinkable indignities and hardships, even the few who do manage to monopolize resources will be condemned to a veritable hell, as they stand bereft of the small innocent pleasures of life, the security and the warmth of their community, and the assurance of a family bond.

The basic question then, even before those who represent privileged groups at the Summit, is how long and how far can you afford to ignore and barter away the human face of existence. Such basic human values cannot be taught through lectures and books, nor can they be nurtured in formal systems which at best treat them as naive and irrelevant aspirations. Such values can only be imbibed in human institutions—small face-to-face communities and families where they are assiduously practiced and lovingly cultivated. We must caution you that this great heritage of mankind can be lost to posterity even if one generation trips and thus causes the chain to be broken. Are we prepared for that cataclysm?

Time is of the essence, friends. We, the disinherited of the earth, particularly in India, wish to make our position clear. The tide of "development" which started rising with the industrial revolution and gained huge momentum during the colonial phase of human history, has now run its full course. The allocation of benefits and costs of this development have been oppressively unfair and iniquitous. The more profitable and amenable activities at every stage have been reserved for themselves by the captains of development—the *Brahmins* (the highest caste) of the new order. The drudgery and the sloth was passed over to the *shudras* (outcasts) comprising the rest. Thus, the creation of a Third World was a precondition of your model. And a Fourth World is in the making, now that the Third World countries have accepted your prescription for their economies. This is the cold logic that must sit in the many minds that deliberate ways and means to save the world. The tide, thus, has reached the furthest shore and has begun to turn menacingly inward. The machine must now feed on itself.

We in Dhorkatta, Bastar, Madhya Pradesh, India are a fragment of this newly created Fourth World. As a logical unfolding of your paradigm, the modern economy of our country, a mere extension of the Western system, has misappropriated our resources. On the principle that you cannot make an omelet without cracking an egg, our little world must disintegrate. It cannot be allowed, of course, to stake any claim to the fruit enjoyed by the estate managers. We either get absorbed in the more powerful system, to the extent possible, or get exhumed and expelled. This logic, if accepted, will not remain circumscribed to one area like ours. It will inform all the disinherited of the Third World and also the deprived of the First and the Second Worlds. The prevailing conditions in the erstwhile Eastern Block and among the non-white minorities in the U.S. and Europe are clear pointers in this direction.

We cannot possibly accept these inevitable consequences of your paradigm as our ordained fate. We do not believe in any iron laws of history, or of economics—free, planned, or mixed in any hue. Man is the maker of history and can chart his own path. Accordingly, after careful consideration, we have rejected outright your paradigm, and its associated lifestyle. It is not only socially unjust, but ecologically unsustainable, besides being devoid of human concerns.

A new paradigm—ecologically viable, socially equitable, and rich in human content—is the historical need of our time. You, at the Summit, have missed the human element totally and considered the social issue only superficially. The outcome of your deliberations will therefore be biased and slanted—perversions which we will have to carefully guard ourselves against. In rejecting your paradigm and raising these issues about the Summit, we are not alone. We echo the deepest feelings of ordinary people across the globe. In doing so, we unwittingly accept a historic role for ourselves, which so far you have refused even to consider. But we are, for all the reasons enumerated above, perhaps better placed in this regard, for we in our system still rank human concerns high. As you can see we have questioned and rejected some of the most fundamental elements of your paradigm. The quality of life cannot be measured by how much we consume or how much energy we utilize. It must, instead, be defined in terms of personal accomplishment of individuals, and the richness of interpersonal relationships within the family and community. A precondition naturally is the fulfillment of basic physical needs for a reasonable living. Accepting this should be the first decisive step towards dismantling the unbearable burden created in the name of so-called development at the cost of earth's fragile ecology. Obviously, human concerns and relationships are non-negotiable. The scope of market, on the other hand, must be circumscribed to the bare minimum. Some areas of life such as enjoyment of leisure must be out of bounds for market, in the interest of a sane society recreating conditions for absorbing dialogue and spontaneous laughter.

Contrary to what the ignorant believe of us, we heartily celebrate advances in science and the expanding horizons of man's universe. But we reject technological regimes built up with an eye on centralization of economic and political power. Technology in such hands has "de-skilled" humans and pushed us from the center to the periphery of the stage. While drudgery can and should be erased through harnessing of technology, it must be remembered that honest physical labor is an essential condition of human life and happiness. In this scheme of things production must be non-centralized in units of human dimension, keeping the master-labor relations to the minimum and slashing heavily on trade, advertisements, and transport. These are the devices of distribution wielded by the haves, whose burden our earth can no longer carry. These are clearly wasteful luxuries created as a sequel to a massive usurpation spree. We reject the production system which has depleted even our non-renewable capital resource-base (subsoil water) for frivolous, temporary gains. By casting this heavy burden on ecology such resources have been rendered out of the reach of ordinary people, forever. Thus, not only do we reject the perceptions and the paradigm, but also the legal framework of the estate-managers which seeks to legitimatize wanton destruction of natural resources and prey even on tomorrow's children of nature.

It should be clear that we are not for the negation of life and progress. What we insist on is that development must have a human face, or else it is tantamount to destruction. Towards this end we wish to announce that a beginning has already been made here in our small corner of the globe. We are clear about our goals, our rights, and our responsibilities. We are establishing village republics (*Nate-na-raj*) in the true spirit of democracy, equity, and fraternity following Gandhian tenets to the extent possible. Our village-republics are not islands in the wilderness, but they encompass even the smallest amongst the ever-expanding circles of the human canvass. We believe that life and vivacity in its totality can be perceived, experienced, and realized only in the microcosms of community and family. It is the community and community alone—not the formal state—which can save the earth for humankind and other forms of life.

So, friends, we have taken upon ourselves a great challenge, with humility yet fully cognizant of the historic role we are playing in one of the most bewildering eras of history. We do not await the advent of a messiah or the conclusion of a revolution—white or red—to move ahead and achieve our goal. The radical structural change associated with the formation of village-republics is a concomitant of the people's struggle. A corollary objective is to assert their will and right of self-governance in the short run and work for a new world order based on equity, fraternity, and democratic values in the long run.

We may, of course, appear momentarily to be moving against the current of history. But that is what it is. We have made a conscious choice that way. But it should be noted, and noted well, that the tide has changed its course. We, therefore, call upon the nations of the world to acknowledge this change, break from the past, and chart out a new path at the Summit for the establishment of a more humane, sustainable, and equitable world.

—31—
Ecological Aesthetics

Michael Tobias

For at least 2,500 years, philosophers have pointed to the impact of our species upon the earth. Plato and Lucretius, for example, were well informed on topics ranging from deforestation to urban overcrowding. Thales understood the perils of water pollution, while Marcus Aurelius, Hippocrates, Epictetus, and others fervently appreciated the philosophy of nonviolence. But what about the planet's impact upon Homo sapiens? Addressing this most obvious and comprehensive connection, molecular biology, demographics, human geography, and even, to a lesser extent, climatology, have elucidated those aspects of human nature and culture forged in the crucible of evolution and the earth sciences. Goethe, Kant, Alfred Wallace, Alexander von Humboldt, and Thomas Malthus are only a few of the more obvious magnets of learning in this regard. Beyond the borders of scientific causality, however, emerges what must be construed as the most fertile and expansive relationship of all: our deeply creative promptings and intrinsically aesthetic responses to nature.

The biological role of beauty has not gone unexamined, though its neuropsychological implications are quantified at the cost of tedium and awkwardness, much like Laban notation (choreography by numbers) or the metrical analysis of G. M. Hopkins. But there are other revealing ways as well to adduce the compelling range of emotions and perception elicited in humans by the earth. The relationship between any planet and its species will afford myriad lessons in reciprocity, of giving and appreciating. The participles themselves are biological, poetical fusion that is alive, and in whose song cycles and primal urgings can be deciphered all of life's generative and restorative properties: the replication of mineral crystals four billion years ago, or the multiple refrains in Handel's *Messiah*.

In a century of unprecedented turmoil, it is more crucial than ever that we understand the power of our connection to earth. Its urgings have become manifest in the guise of our potent and astonishingly subtle expressions. Humanity has channeled the forces of feast and famine, sea-change and starry nights, imbalance and brutality, perfection and disarray. It has been suggested that nothing is safe in this world that is not absolutely sacred. Art sanctifies, occasional bombings at museums aside. Because our grandest artistic ideals have been vouchsafed, culturally agreed upon, and glorified, we may perhaps seek in those aesthetic impulses and injunctions a global strategy of overwhelming importance. What the Greeks termed *physiolatry*, the love of nature, may hold the key to our success or failure as a species. So thoroughly have the natural splendors of earth utterly guided our passions, visions, dreams, metaphors, apperception, self-consciousness, and idealisms—the neural network of the psyche—that it appears reasonable to suggest that in our art we have voiced this indebtedness. Moreover, if art is any window at all on the psyche, then it is clear that more than a powerful syllogism is at work here. In the illuminations of art, we obtain clues to a biological link that is our solace and preservation. In these times, the life of every individual, of all species, hinges upon this crucial reciprocity, this connection between nature and psyche, creation and appreciation. The biological earth's very survival is now dependent like never before upon our ability to celebrate beauty, to respect nature with all our hearts.

Such *natural positivism*, an anodyne of universal dimensions, constitutes a more far-reaching and challenging initiative than mere problem-solving, the daily duty to which economists, engineers, and politicians attend. Natural positivism, as I would imbue the phrase, speaks to a broader affirmative stance, an aesthetic perception of nature which holds that life is endlessly rich, endlessly original, a miracle at the beginning, middle, end, middle, and beginning. Such a position has been articulated unambiguously over many millennia. One could cite a particularly poignant burial at the Shanidar Caves in the Zagros Mountains sixty thousand years ago, a disinterred skeleton found to have been buried with a tiara of hollyhocks and marigolds; or the sensuous depicture of a stag at Lascaux forty thousand years later. Human reverence for nature may be fraught with infuriating contradiction in the real world—the official animal of California, the grizzly bear, has not been seen in the state since the 1920s; and while India worships the Ganges, it has largely killed large sections of the river—but its aesthetic counterpart knows very little ambivalence. Natural positivism has its own ecological continuity and invariants, just as the gene, the cell, and any ecosystem are possessed of life-stabilizing population dynamics (such as rookery size, grazing range, feeding and mating habits.) Indeed, it has achieved its biological equivalencies in the very artistic impulses that define Homo sapiens. To create, ergo, to be. (This does not mean, however, that there are any predictable rules: the young Hitler, for example, was a copious if lousy landscape painter.)

This "equation" of biological equivalency does suggest, however, endless human itineraries and lines of inquiry, all valid, all fundamental to the study of ecological aesthetics. So prolix is this field that every artistic nuance can have deeper naturalistic meaning, analogy, and import. "Real presences" (to use George Steiner's phrase) abound, the visible world becomes haunted by the connections as semblances, simile, and transmetaphors proliferate. The working vocabulary of human experience takes on additional artistic weight, as it suddenly reveals urban landscapes and mountain ranges, both; plains of heaven, churning seas, utopian fancies, or the more austere aesthetics, driven equally by a vision of nature, inherent, say, to Bauhaus or Japanese furniture. This is not simply the psychology of art. Ecological aesthetics summons a penetrating profile of the human being in the guise of his or her artistic passions; a profile that is not merely of the psyche, but of the total organism, as scientists are wont to describe it. For what is nature but nature inside us, and all around us. Thus, individual preferences, idiosyncrasies, elocution, and style accumulate into patterns of behavior, the collective unconscious, cities, neighborhoods, cultural norms, whole civilizations. Just as other historians (i.e., Vico, Gibbon, Toynbee) have tracked the rise and fall of civilizations, students of ecological aesthetics may begin to perceive anew the mechanisms of conflict resolution or economic disparity. Natural positivism, a phrase I am using as synonymous with ecological aesthetics, may intone bucolic bounty, or suggest doing less with less. It offers lessons in international exchange, in sustainable utilization, transgenerational equity, assisted self-reliance, zero economic growth, the legal framework for a global commons and the macro-socialistic ramifications that would entail. In short, ecological aesthetics provides valuable insights into environmental understanding in general. In all those enterprises of human being, none so clearly and decisively speak to our future as those widely hailed to be "natural." This natural positivism reforms our entire understanding of history, of countless joys and downfalls, engineering chimeras, political hazards, military mania, and the odd Renaissance. At times, the connection requires a leap, or a stretch. We are not accustomed to treating the "data" in this way. Ordinarily, we do not "study" a painting or poem. Our goal is simply to enjoy the work, to be inspired by it. Why demand more of it than that, which, as is often the case, was the artist's intention?

My point is not to beg more from the already perfect realm of art, but rather to recognize an important source of unexpected power, a tradition linking seemingly disparate qualities of the human experience. Our aesthetic predilections are not merely subsumed by our biology. It may well be that the aesthetic can mold the biological. That what we take to be implacable, defined by evolution, is in fact an open question; that our artistic ideals can become self-prophesizing, given the flexibility and reason of human personality. Our behavior is not a closed system. Art can greatly modify it. What kind of art, one may ask?

I would cite certain seminal moments in the history of aesthetics. Begin, for example, with those remarkable 154 remaining lines by the Egyptian

Middle Kingdom Twelfth Dynasty poet laureate, Khety, son of Duauf, "The Report about the Dispute of a Man with His Ba." The *ba* referred to the man's soul, which speaks out on behalf of the earth, advising its body to cling to the marvels of life, rather than to be swayed by all of its quotidian inclemencies. The Egyptians likened the ba to the jabiru bird, a now-extinct stork, life-giver, source of renewal in the next life. But Khety asked, why not in this life? He emphasized no immortal journey into the underworld, but rather, the smell of myrrh, the rain-washed path, the sense of drifting down the Nile on a breezy day, returning home after an expedition. "To be adamant about life!" Those were Khety's words, conveyed at a time when his contemporaries were more consumed by the notion of building monuments, the pyramids of Zoser and Khufu, the royal tombs at Tanis, the temple of Rameses III at Medinet Habu, the equivalent of today's space shuttle and skyscrapers. Yet, in this one poet, whose manuscript was discovered in a bundle of Egyptian documents in 1843, is an exquisite ecological code of ethics that prefigures Jacopo Sannazaro's *Arcadia* and Walt Whitman's *Leaves of Grass* by more than three thousand years.

In the early sixth century B.C., Mahavira, the 24th Tirthankara of Jainism, conveyed his own ethical revolution, totally rooted in natural science. He wandered for forty years, not through the deserts of Sinai, but India's tropics, taking an inventory of nature that would amount to literally millions of species. By comparison, a century later, Aristotle could find less that six hundred species to write about. Mahavira rhapsodized nature in a strict language given to the ascetic, but translated those poetic impulses into an ecological code of behavior which, more recently, people like St. Francis, Gandhi, even Mikhail Gorbachev have called for.

Mahavira, and the Jains, provide a blueprint for ecology that is at once artistic, compassionate, and problem-solving. Mahavira's message of ahimsa, of nonviolence, brilliantly highlights what the world has now tragically compressed into a single challenge between North and South, between the rich and the poor, the aggressive and the restrained, between meat-eaters and vegetarians: Live simply so that others may simply live.

Fifteen hundred years after Mahavira, from Tibet, another life comes down to us in the form of a most remarkable literary biography written on Mount Everest in the year 1135 by one Rechung-Dorje-Tagpa, and chronicling the abundant, peripatetic Jetsun Milarepa. The work, translated as *The 100,000 Songs of Milarepa* invokes centuries of introspective heroism in the shadows of Buddha, an Odyssean journey towards nirvana that confronts ignorance, poverty, disease, every adversity, but manages to transmute these woes, these transitory illusions, into glory. Once again, nature—Tibetan outback—becomes the vehicle for promulgating a science of behavior.

In Japan's late fifteenth century, the Ashikaga Shogun Yoshimasa spins the equation, by incorporating the elements of the tea ceremony into his comprehensive vision of a private home. That home, Gin-kaku-ji, today a national monument in Kyoto, boasted of twenty-two species of moss, koi ponds, sand gardens, "Waiting-for-the-Moon-Hill," enormous evergreens, bright maples, and the *chashitsu*, the first tea room in all Japan. While the Onin Civil War raged throughout the country, resulting in hundreds of thousands of deaths, Yoshimasa—having renounced his shogunate—cultivated the art of tea, the "only Asiatic ceremonial which (today) commands universal esteem," wrote the early twentieth century art historian, Kakuzo Okakura. After all the war embers had died out, it was the tea, and the moss, which lived on. Later tea masters, like Rikyu and Oribe, would examine every stone at Gin-kaku-ji in an effort to determine *what is enough*. Oribe prescribed six practical, to four aesthetically placed stones. Rikyu argued for precisely the opposite.

In 1930, the Turin National Library acquired some musical manuscripts from an Italian monastery. Among them were a number of works by one Antonio Vivaldi. Until that time, his name was vaguely associated with the mid-nineteenth century revival of Bach. But suddenly, Vivaldi scholarship exploded. By the early 1970s over 740 works by the master had come to light, the most prominent of them being his eighth opus containing twelve concertos and titled *Il cimento dell-armonia e dell-inventione (Contest between Harmony and Invention)*. The first four of these concertos were called *Le quattro stagioni (The Four Seasons)* first published in 1725. *The Four Seasons* stands out, once again, as a pivotal moment in the human psyche: that musical summit where the human spirit and its earth mother coalesce in naked, unblushing unison; in music whose calligraphy, according to musicologist Marc Pincherele—Vivaldi's seminal biographer—resembled "certain rough drafts by Rimsky-Korsakoff, where the stems of the sixteenth notes bend like a field of wheat before the storm." Vivaldi's half-a-dozen "tempests" seem to have been modeled after a predecessor, the Venetian painter Giorgione, whose own unprecedented *Tempesta* Vivaldi probably studied at the home of Messer Gabriel Vendramin, a young and well-mannered descendant of a great late-fifteenth-century doge. In Giorgione's day, all of Europe was warring against Venice. Yet the young painter who was soon to die of plague in his lover's arms decided to focus on the perfectibility of love, of human nature, of nature itself, rather than on the converging darkness. He did so—as would Vivaldi more than two centuries later—by highlighting the storm and offering an icon of eternal human renewal. Together, Giorgione and Vivaldi conveyed a potent tapestry of mauve and umber glazes, of spring, summer, fall, and winter, of lust and oxygen, animal cries and howling winds, dark chords, soft laments, urgent tempos, and evening pianissimos. It was a revolutionary language of nature, both in paint and in song, that spanned the Italian Renaissance and continues to school posterity not in political science, but the human potential. Both Giorgione and Vivaldi are offering us clear signals, solutions to our ecological woes. These are not overtly monetary resolutions, but spiritual ones, requiring patience, love, astute observation. It's all there for the senses to grasp, beyond the recondite or specialized. A universal language.

Similarly, Jan Van Eyck, in his *Bethrothal of the Arnolfini* (1434, at the National Gallery in London) scales his own Mount Everest, in an ordinary Flemish room. "For the first time in history," writes art historian E. H. Gombrich, "the artist became the perfect eye-witness in the truest sense of the term." In an age obsessed with maps and exploration, with mercantile conquest, Van Eyck charted a course to the inner light, where a couple touches hands, and love blossoms, and every detail of life becomes ecological, navigational, organic, a crucial part of experience. Remove one detail, and the whole balance collapses.

The interdependency of Van Eyck's Brugge—that serene center of the early Renaissance—presages an even greater ecological aesthetic in the work of Johannis Vermeer of Delft. While Vermeer's neighbor, the amateur microscopist Antony van Leeuwenhoek was pushing the optical frontier, discovering a larger population of organisms in his own mouth that the total number of citizens of the United Netherlands combined, Vermeer was capturing the very light which made scientific revelation possible. It is said that Vermeer conquered light the way a bird conquers gravity. Vermeer's two known landscapes, *View of Delft* and *Street in Delft*, declare the profound preeminence of landscape contemplation, of perception that is forever beholden to the interior and continuous nature of being. To fashion this wilderness metaphysic within the world of the familiar, Vermeer layered his paint like gossamer, like strands of gold silk, burying the natural juxtaposition of hues and tonality in a constant haze of approaching invisibility so that, unexpectedly, without the slightest overture, something happens that we cannot explain, in every object, in the air, in our eyes. It is this phenomenon of absolute stillness in the heat of reverie that perpetuates the utter mystery that was Vermeer. His observational powers were astounding. By renouncing the grander scale, Vermeer discovered human scale. For the Dutch painter, small was beautiful.

It has been said that a door handle, a bed, a pair of shoes, and the height of happiness, namely, of a human being, rarely if ever change. A bed remains the same size, more or less. And a person a person. By focusing on human scale in his work, Vermeer, like Van Eyck before, added other dimensions to the realm of ecological aesthetics.

In England, Savoy, and Italy, the poet Percy Shelley furthered this personal naturalism. By July 1816, Mont Blanc had been climbed eleven times and had seen many other attempts. Yet Shelley chose to ignore the information in his creation of a mountain that was virgin, undiscovered, a paradise of human yearning and poetic completion. He wrote of his poem, "It was composed under the immediate impression of the deep and powerful feelings excited by the objects which it attempts to describe; and as an undisciplined overflowing of the soul, rests its claim to approbation on an attempt to imitate the untamable wildness and inaccessible solemnity from which those feelings sprang." And it was this sensibility that prompted Shelley, in his poem "Mont Blanc," to speak of—

> The secret strength of things,
> Which governs thought, and to the
> infinite dome
> Of heaven is a law, inhabits thee!
> And what were thou, and earth, and
> stars, and sea,
> If to the human mind's imaginings
> Silence and solitude were vacancy?

Cryptic, unresolved, intimating the complete apprehension of nature by the human soul, Shelley's identification with wilderness was a modern one, truant, tested, on the lam. Shelley decried, as did Wordsworth and Dickens, the coming onslaught of industrialization. Among countless nineteenth century musical and pictorial talents, this machine culture portended of apocalypse, of ecological existentialism. The American founder of the Hudson River School, Thomas Cole, first detected clues to it in the ever increasing tree stumps peppering the American landscape. By the turn of the century, the New Jersey painter, George Inness, depicted the sunset on American nature, his works deeply stirring, empathetic, but aware of awful destiny. Inness, unlike his Hudson River School forebears who were content to idealize grandeur, sought harmony in discord, already too aware of what was happening to the wilderness. What distinguishes Shelley at the beginning of the nineteenth century from Inness at its conclusion, is not a changed soul, but a new orientation. Shelley, who of course died at the tender age of twenty-nine, did not live to see the "invention" of Manhattan, just a few hours by train from Inness's farm in Montclair.

With the machine came Cubism, the fracturation of the world, the breaking down of impregnable substance after a long, untouchable history from Democritus to Dalton. J. J. Thompson's discovery of the electron in

1897, the first acquaintance with half-lives, chrono photography, the strobe, the motion picture, gave over a new world view, all in a cache of fragments, of absurdity and, conversely, the power of realism. World War I, and ten million dead, along with the 1918 influenza pandemic which wiped out another twenty million, furthered the fragmentation of nature in the human mind. Nature, ultimately, was reduced to a form of "nausea" in Sartre's classic by the same title, published in 1938, an unwitting prelude to the Holocaust. Sickened by his connections to earth, unable to flee the onrush of ecological awkwardness, Sartre's antihero, Antoine Roquentin, goes berserk, is literally crushed from within by the inability to cope, and this inverted frenzy, this twentieth-century vision of despair, Sartre aptly named and exorcised during Roquentin's famed museum tour. "Farewell, lovely lilies, elegantly enshrined in you painted sanctuaries, good-by, lovely lilies, our pride and our reason for living! Good-by, you bastards."

The aesthetic journey confronting these moodswings and revelations is an ecological one, in the sense that the human organism is constantly modifying its behavior according to the environment. In no other behavioral realm does an organism achieve such subtle response and modification as in its dreams and ideals which, in the case of *Homo sapiens*, takes on a bewildering array of expressions. Much has been made of the dances of bees, the natural hybridization of plants, the colors of avian plumage, and the songs of the whales. We know next to nothing about any other species (the most closely watched organisms are domestic dogs and bacterium E. coli). There may be as many as one hundred million species on the planet (or there were, at the turn of the century) but there are fewer than 1.6 million labeled by science today. And usually, a label means just that—a name and nothing more. Nothing about the organism's behavior, shape, locale . . . nothing! We are still at the beginning of our own evolution, yet have already declared our self-destructive intentions.

Psychologists agree that when a young person continually threatens to commit suicide, that person should be taken seriously. Similarly, *Homo sapiens* as a species has been sounding another type of alarm, a wake-up call; in essence, we have been saying something that is crucial to our own survival, and it is to be found in the very art which we love and treasure. This natural positivism may be the most enduring and life-transforming link in all of the behavioral sciences between nature and the human organism. Our past, as well as our future are encompassed by the laws of form, substance, and inclination inherent to its affirmations and urgings. The serious study of ecological aesthetics might offer us the tactical clues and inner peace we desperately need if we are to reinvest the world with the beauty and goodness we are capable of perceiving there.

The Bradley Method of Bush Regeneration

John Seed

A method has been developed in Australia for regenerating native bush. It is named the Bradley Method after the two sisters, now deceased, who devised it.

Should it be our wish to bring back the native vegetation that once covered a particular piece of earth, then, they found, no heroic tree planting measures are called for. Rather, this humble technique requires us only to remove all foreign influences while causing the minimum possible disturbance to whatever native vegetation still exists.

Thus, the first step may be to fence off the area we have chosen, to keep cows or goats at bay. It may also be necessary to take steps to prevent fire from invading the land. We must then be able to identify all species of plants that we encounter, both the exotics and those native to the area. We need to recognize them not only in their mature form but also when their seedlings first poke out of the ground. *Then* the method is simple: remove the exotics without treading on the natives. Encouraged in this way, the native species begin to come back, growing stronger in each ensuing season.

There is only one other rule: start from the strength. It may be that in the area we wish to heal there are deep scars, erosion gullies perhaps, that break our heart, and it is our wish to immediately tend to these. In order to succeed however, we must resist this temptation and start from the strongest expression of native vegetation in our management area.

If our area is an inner-city park that has been lawn for a century, our beginnings may be from a tiny patch that the mower couldn't reach and where a few native weeds flourish. If there is a forgotten corner where a few pioneer tree seedlings have emerged, we start from there. Carefully stepping backwards, removing exotics as we go, we invite the bush to follow. It is painstaking work. Each year, the process accelerates as the native intelligence of the place emerges and the life-force quickens.

More and more species emerge as the conditions necessary for their growth are recreated. As one species of pioneer trees completes its work in say, repairing the soil with shade and leaf mulch, its numbers become fewer and are succeeded by the next generation. The microclimate slowly changes and, one day, after perhaps seven years of this patient, rewarding service, we may find to our astonishment, a seedling emerging of a climax species that has not been seen here in the city for one hundred years. Was the seed dropped by a bird that alighted in the branches of a pioneer tree now reaching one hundred feet above? Is it possible that the seed lay dormant in the ground since it was first cleared, waiting for this moment when conditions were again suitable for its return? We will never know.

And when the accelerating advance of the native bush finally reaches that erosion gully, it now has the vigor and the necessary species to be able to recolonize, integrate, and slowly come back into harmony.

Human is not the hero, proudly planting thousands of trees, reclaiming the desert, healing the earth. Rather we are humble in the face of the super-human intelligence of nature, and *invite* the original nature of the place to return.

There is something very spiritual about the Bradley Method. Encoded within it is a deep trust in the native intelligence of the earth—she *knows* what is meant to grow in this place and she also knows, unerringly, the particular stages of succession that will best take us from whatever kind of degradation exists at present back to climax.

In my travels, I have encountered systems akin to the Bradley Method in several different countries. I found one example as I traveled around India in 1987 lecturing on rainforest conservation and deep ecology. In Bhopal I visited one of the most enlightened foresters I have ever met. His name was Chaturdevi and he was professor of the new school of forestry that had been established in that city just a few years before. His school had been granted a large area of ground, a couple of thousand acres as I recall, and the first thing he had done, before the first brick was laid for the school buildings, was to fence the land. It was at that time a desolate thorny desert denuded by goats and recurrent fires set by the goat-herders to encourage succulent new growth. Chaturdevi hired armed guards to keep these at bay. The first task he asked of his students was to inventory the vegetation that grew there. In the beginning, they discovered stunted remnants of a few tree species which had managed to survive the former regime—just a few sticks here and there whose leaves had been chewed back as soon as they emerged.

By the time that I saw the land some four years later, more than eighty species had re-emerged as I recall, and in many places, the vegetation was pumping, accelerating back towards its climax status. The armed guards were still there.

In many other places, from Russia to the United States, I have found, to my surprise, understandings akin to the Bradley Method emerging independently and unbeknownst to each other. Perhaps it should come as no surprise. Perhaps this phenomenon is *itself* a manifestation of the Bradley Method, only working here on the level of the human psyche rather than the biology of a landscape. Surely the human psyche is itself a product of the landscape—we ourselves grow from the soil, are made of soil. In this case, the most appropriate metaphors for understanding psyche are biological ones. Techniques that facilitate the return of native vegetation may also help us understand how wild common sense can return to the denuded mind. The spontaneous emergence of "the Bradley Method" in different places around the world can then be seen as an expression of the return of a trust in nature after centuries and millennia of human arrogance.

What I first learned of by the name "Bradley Method" then may be one stage in the succession of the return of native wisdom and humility to the clear-cut modern mind, when the exotic influences of anthropocentrism are removed. By anthropocentrism I mean the ubiquitous modern idea that the human is the center of everything and that order comes into the world only through human control and ingenuity. The rejection of anthropocentrism has sometimes been termed "deep ecology."

One of the understandings of deep ecology is that the sense shared by most modern humans, of being isolated, alienated, and separate from nature is illusory. In fact, we are Earthlings, we *belong* here. We have evolved on this planet for four billion years of organic life and are made of earth. Our soul too, our psyche, is earth-born, emerging from the exquisite biology of this planet, continuous with it. The ubiquitous illusion of separation springs from the false ideas of human "otherness" and superiority that thousands of years of Judeo-Christian and other traditions have created within us.

As we root out these pernicious false ideas of our own grandeur and importance, we "fall in love outwards" (Jeffers) and the truth spontaneously emerges of who we *really* are, "plain members of the biota" (Leopoid). As the exotic influence of the dominant paradigm recedes, we realize (with Commoner) that "nature knows best" and our native intelligence pops spontaneously from the ground of our being.

When we see the Bradley Method as being equally applicable in the reawakening of native human intelligence, as in the re-emergence of a biological ecosystem, several corollaries suggest themselves. Firstly, we don't need to plant new ideas in each others minds. If we can root out the alien ideas, the ecological insight springs forth spontaneously. We need to know ourselves, to create an inventory of our mind, to learn to recognize the ideas, feelings, habits, blockages which prevent us from experiencing our unity with nature. Which parts of us are harmonious with our larger system?

How can we compassionately root out destructive habits and conditioned ideas without unnecessary disturbance and self-hatred?

Secondly, start with the strength: there's nothing wrong with preaching to the choir. In fact it is more important to strengthen the experience of deep ecology among those who already love nature and work for the earth than to waste our energy trying to convince the CEO of the EXXON erosion gully about the importance of a biocentric ethic. Strengthening ecological empathy and insights within the conservation community will make it ever stronger and more capable of making inroads into corporate culture, the Vatican, and other bastions of anthropocentrism.

Endnotes

Chapter 7 The Idea of the North

References

Anderson, Laurie. *Tales from the Nerve Bible*. In press.

Gould, Glenn. *Glenn Gould Reader*. Edited by Tim Page. New York: Knopf, 1984.

Gould, Glenn. "The Idea of North." Radio program. Toronto: CBC Radio, 1967.

Helprin, Mark. *Winter's Tale*. New York: Bantam, 1984.

Kaysen, Susanna. *Far Afield*. New York: Vintage 1988.

Kpomassie, Tété-Michel. *An African in Greenland*. New York: Knopf, 1981.

Lopez, Barry. *Arctic Dreams*. New York: Macmillan, 1986.

Millman, Lawrence. *Last Places*. Boston: Houghton Mifflin, 1991.

Nelson, Richard. *The Island Within*. San Francisco: North Point Press, 1990.

Rasmussen, Knud. *Across Artic America: Narrative of the Fifth Thule Expedition*. 1927. Reprint. Westport, Conn.: Greenwood, 1970.

Reed, Peter, and David Rothernberg, eds. *Wisdom in the Open Air*. Minneapolis: Univ. of Minnesota Press, 1992.

Schafer, R. Murray. *Music in the Cold*. Toronto: Arcana Editions, 1977.

Shelley, Mary. *Frankenstein*. 1818. Reprint. London: Everyman's Library, 1992.

Snyder, Gary. *The Practice of the Wild*. Berkeley: North Point Press, 1990. Pp. 73–74.

Vollman, William. *The Ice Shirt*. New York: Viking, 1990.

Chapter 12 Understanding the Great Mystery

1   The link between Thoreau and Native thought has been explored by Professor George Cornell, an Ojibway writer from Michigan. Cornell further develops, in his unpublished Ph.D. thesis, the strong links between the whole idea of a conservation ethic and Native American traditions. That link is not the subject of this essay, so I will only mention it in passing and urge those interested in pursuing it further to seek out Cornell's writing, some of which has appeared in *Akwekon Journal*, a Native American publication.

2   Long Standing Bear Chief (Harold Grey), *Ni Kwo Ko Wa* (Browning, Montana: Spirit Talk Press, 1992).

Chapter 18 India's Earth Consciousness

1   Buddhaghosa, *The Path of Purification (Visuddhimagga)*, translated from the Pali by Ghikkhu Nyanamoli (Boulder: Shambhala, 1976), IV: 24–26, pp. 127–29.

2   Ibid., IV:29, pp. 129–30.

3 Ibid., V:3, p. 177.

4 Ibid., V:7, p. 178.

5 Ibid., V:9, p. 179.

6 Ibid., V:21, p. 181.

7 Ibid., IV:31, p. 130.

8 Ann Spanel, "Interview with Vandana Shiva," in *Women of Power* 9 (1988): 27–31.

Chapter 19 Spirit in Action

1 William Bridges, *Surviving Corporate Transition* (New York: Doubleday, 1988).

Chapter 22 Tribalism

1 Jacques Ellul, *Perspectives on Our Age* (New York: The Seabury Press, 1989), 70–71.

2 Herbert Schneidau, *Sacred Discontent: The Bible and Western Tradition* (Baton Rouge: Louisiana State Univ. Press, 1977), 39.

3 Thomas Foster, "Amish Society: A Relic of the Past Could Become a Model for the Future," *The Futurist* (December 1981).

4 Peter Berger and Hansfried Kellner, *The Homeless Mind* (New York: Random House, 1974).

5 Also the notion of "development" as an extension of "The West." Gustavo Esteva saves his sharpest criticism for this in "Regenerating People's Space," *Alternatives* 12: 125–152.

6 Michael Ignatieff, *The Needs of Strangers* (New York: Viking, 1985), 29.

7 *Mishna Kiddushin* (Jerusalem: Weinfeld, 1985 ed.), 1:9–10J.

8 Isaiah 24:4–5.

9 Ibid.

10 Robert Bellah, *Habits of the Heart* (Berkeley: University of California Press, 1985), 114.

11 Ibid., 16.

12 Russell Means, "The Same Old Song," in *Marxism and Native Americans*, edited by Ward Churchill (South End Press, n.d.), 29.

13 Naomi Pasachoff, *Basic Judaism for Young People* (New York: Behrman House, 1986), 69.

14 Max Dimont. *The Amazing Adventure of the Jewish People* (New York: Behrman House, 1984), xiv.

15 Churchill, op cit., 29.

16 *Mishna*, Avot 1:1.

17 Jack Waddell and O. Watson, eds., *The Urban Indian American in Urban Society* (Boston: Little, Brown, 1971), 207–42.

18 U.S. policy changed in the late nineteenth century. At this point assimilationist policy was officially implemented on a massive scale. This included allotment of

tribal land to individuals, accelerated western education, and sending kids off the reservation. The culmination was probably the General Allotment Act of 1887.

19 Deuteronomy 11:12.

20 *Mishna Pirkei Avot* (New York: Hebrew Publishing), 194.

21 Ecclesiastes 3:19.

22 *Genesis Rabba*, 10:8.

23 *Tanna Debei Eliyahu Rabba*, 2.

24 *Exodus Rabba*, 31, 15.

25 See the writing of G. Hobson and Paula Gunn Allen, *The Sacred Hoop* (Boston: Beacon Press, 1986).

26 *Yalkut Shimoni*, Psalms, 104.

Chapter 23 The Sacred Womb

1 Carl Jung and Carl Kerenyi, *Essays on a Science of Mythology* (Princeton: Bollingen Series/Princeton. 1973), 43.

References

Campbell, Joseph. *The Mythic Image*. Bollingen Series C. Princeton, N. J.: Princeton Univ. Press, 1974. Pp. 217, 237, 238, 255, 262–265, 270.

Getty, Adele. *Goddess*. New York: Thames & Hudson, 1990. Pp. 43, 66.

Gimbutas, Marija. *The Civilization of the Goddess*. San Francisco: HarperCollins, 1991. Pp. 222–223, 226, 228, 262, 265, 286, 288, 290, 292.

Gimbutas, Marija. *The Language of the Goddess*. San Francisco: HarperCollins, 1989. Pp. 149–151.

Miracle of Life, The. Videotape. New York: Crown Video/WGBH Boston, 1986.

Neumann, Erich. *The Great Mother*. New York: Bollingen Foundation, Series 47, 1963.

Scher, Bob. "Teotihuacan: City of the Gods." *Parabola* 18:4 (November, 1993): 91.

Walker, Barbara. *The Woman's Encyclopedia of Myths and Secrets*. New York: Harper & Row, 1983. P. 740.

Chapter 27 Creating an Ecological Economy

1 Barbara Ward and Rene Dubos, *Only One Earth: The Care and Maintenance of a Small Planet* (New York: Norton, 1972), 220.

2 Quoted in Bill Devall and George Sessions, *Deep Ecology* (Salt Lake City: G. M. Smith, 1985).

3 Arne Naess, *Ecology, Community and Lifestyle: Outline of an Ecosophy* (New York: Cambridge University Press, 1989).

4 E. F. Schumacher, *Small Is Beautiful* (New York: Harper & Row, 1973).

5 Ben Jackson, *Poverty and the Planet: A Question of Survival* (Harmondsworth: Penguin, 1990).

6   *The New Internationalist.*

7   Susan George, *A Fate Worse Than Debt* (New York: Grove Press, 1988).

8   Vandana Shiva, *Staying Alive* (London: Zed Books, 1989).

9   *Breakthrough Magazine: Journal of the Prairie Fire Organizing Committee*, Vol. 15 (San Francisco: John Brown Book Club).

10 Wendell Berry, *Home Economics* (San Francisco: North Point Press, 1987), 110.

11 *State of the World.*

Chapter 29 The Age of Light

1   See, e.g., Robert Theobald, *The Rapids of Change* (Indianapolis: Knowledge Systems, Inc., 1987).

2   Jeremy Rifkin, *Algeny* (New York: Viking, 1983), and Jeremy Rifkin and Ted Howard, *Who Shall Play God?* (1977).

3   Jeremy Rifkin, *Biosphere Politics* (New York: Crown, 1991), chap. 9, p. 65.

4   Gar Smith, *Patently Ridiculous* (Berkeley, California [P.O. Box 27] 94701, 1980).

5   "Gene Amplification," *Economist* (July 27, 1991): 76.

6   *Business Week* (June 3, 1991).

7   "The New Alchemy," *Business Week* (July 29, 1991).

8   Eric K. Drexler, *Engines of Creation* (New York: Doubleday, 1986).

9   "Creating Chips an Atom at a Time," *Business Week* (July 29, 1991): 54.

10 Ilya Prigogine, *From Being to Becoming* (San Francisco: HarperCollins, 1980).

11 Arthur M. Young, *The Reflexive Universe* (New York: Delacorte, 1976).

12 Riane Eisler, *The Chalice and the Blade* (San Francisco: HarperCollins, 1980).

13 Nigel Calder, *The Green Machines* (New York: G. P. Putnam & Sons, 1986).

14 Gregory Bateson, *Mind and Nature: A Necessary Unity* (New York: Bantam Books, 1980).

15 Hazel Henderson, *The Politics of the Solar Age* (Indianapolis: Knowledge Systems, 1988) [current ed.], 225 on Frederick Soddy.

16 *U.S. MAB Bulletin*, Vol. 15, #3 (August 1991), U.S. Department of State, Washington, D.C. 20522-3706.

17 Allen V. Kneese, "The Economics of Natural Resources," in *Population and Resources in Western Intellectual Traditions*, a supplement of *Population and Development Review* 14 (1988).

18 World Resources Institute, "Minding the Carbon Store: Weighing U.S. Forestry Strategies to Slow Global Warming," Mark C. Trexler, January 1991.

19 Another classic resource for further exploration of these understandings is the four-volume *Dynamics: the Geometry of Behavior* by Ralph H. Abraham and Christopher D. Shaw (Santa Cruz: Aerial Press 1982–88).

—Contributors—

Angeles Arrien is a cultural anthropologist, award-winning author, educator, and corporate consultant. She lectures internationally and conducts workshops that bridge cultural anthropology, psychology, and comparative religions. She teaches the universal components of communication, leadership skills, education, and health care. She is also the author of *Signs of Life*.

Rick Bass is the author of several books of natural history, including *Winter* and *The Ninemile Wolves*. He is working on a novel, *Where the Sea Used to Be*, and a book of non-fiction about the search for grizzlies in Colorado. He lives in Montana.

Wendell Berry lives and farms in Henry County, Kentucky. He is a poet and novelist as well as an essayist. His most recent books are *Fidelity*; *Sex, Economy, Freedom and Community*; *Entries*; and *Watch with Me*.

Joseph Bruchac is a writer and storyteller of Native American and European ancestry and a registered member of the Abenaki Nation of Vermont. A graduate of Cornell University, where he majored in English and minored in wildlife conservation, his most recent books include the novel *Dawn Land*, and two children's books, *The First Strawberries* and *Fox Song*.

Christopher Key Chapple earned his doctorate in the history of religions and theology at Fordham University and trained for several years in classical Yoga at Yoga Anand Ashram. He is currently professor of theological studies at Loyola Marymount University in Los Angeles and is author of *Karma and Creativity* and *Nonviolence to Animals, Earth, and Self in Asian Traditions*, as well as editor of *Ecological Prospects*.

James Cowan is the author of some fifteen books, including *Messengers of the Gods*; *Letters from a Wild State*; and *Mysteries of the Dreaming*. Cowan has spent much of his life exploring the world of traditional peoples, such as the Berbers of Morocco, the Tuareg of the Central Sahara, and the Iban of Borneo. He is interested in their way of reconciling environmental exploitation and sustainability through the use of myths, poetry, and ritual.

Avram Davis, Ph.D., is the author of many articles and two books. He runs a meditation and Jewish spirituality network in Berkeley, California, and is director of Chochmat HaLev (Wisdom of the Heart), an independent renewal Bet Midrash, or, Wisdom School. The path of particularism and tribalism forms one of the foundations of his spiritual message, a "path of the passionate heart."

Annie Dillard was only twenty-nine when *Pilgrim at Tinker Creek*, her second book, won a Pulitzer Prize in 1975. Her honors since then have included a National

Endowment for the Arts grant, a Guggenheim Foundation grant, and a National Book Critics' Circle Award nomination in 1987 for *An American Childhood*. Dillard is the author of eight books, including a volume of poetry and one collection of essays.

Gretel Ehrlich's essays have appeared in *The New York Times*, *Harper's*, and *The Atlantic Monthly*. She is the author of *The Solace of Open Spaces*; *Heart Mountain*; *Islands, The Universe, Home*, and a collection of stories, *Drinking Dry Clouds*. She is a Guggenheim fellowship recipient and lives in Wyoming.

Matthew Fox is the author of several books, including *Sheer Joy*; *Creation Spirituality*; and *Liberating Gifts for the Peoples of the Earth*. A Dominican priest, theologian, and educator, he is the founding director of the Institute in Culture and Creation Spirituality at Holy Names College in Oakland, California, where artists, Native peoples, social transformers, scientists, theologians, and psychologists strive together to discover the cosmology that our times are offering us. In December 1988, under pressure from the Vatican, he was silenced by his order for a year.

Thich Nhat Hanh is a Zen master, poet, and peace advocate. In 1967, Martin Luther King, Jr. nominated him for the Nobel Peace Prize, saying "I do not personally know of anyone more worthy of the Nobel Peace Prize than this gentle monk from Vietnam." Author of *Being Peace*; *Peace Is Every Step*; and the *Miracle of Mindfulness*, he lives in a small community in France, where he teaches, writes, gardens, and works to help refugees worldwide.

Linda Hasselstrom has lived in western South Dakota for forty years, earning her living by ranch work, writing poetry and non-fiction, and teaching workshops in writing. She has written, "I work to bring my life into a circle: writing things I can respect, publishing work I respect, laboring at riding, branding, gardening, taking care of the land, and doing it all with an awareness of how those things fit together." She is the author of several books.

Tom Hayden is a senator in the California legislature. He has played an active role in American politics for over three decades. John Kennedy's speech writer said that Hayden "without knowing it, inspired the Great Society." He chaired the campaign for the Big Green initiative and is the author of seven books, including the forthcoming *Towards a Gospel of the Earth*.

Hazel Henderson is an internationally published futurist, lecturer, and consultant to organizations in over thirty countries. Her books inlcude, *Creating Alternative Futures: The End of Economics* and *Politics of the Solar Age: Alternatives to Economics*. She is the Director of Worldwatch Institute and founded the Center for Sustainable Development and Alternative World Futures.

Petra Kelly—hailed as one of the most influential women in Europe in this century—was a lifelong grass roots activist, a leading figure in the peace and human rights movements, and co-founder of the Green Party, which has profoundly influenced world politics. Kelly served in the West German parliament from 1984 to

1990, and was planning to run for the European parliament in 1994. However, she was murdered in the fall of 1992, a crime still unsolved.

Barry Lopez is the author of many books and essays, including *Arctic Dreams*, for which he won the National Book Award for non-fiction. His other works include *Desert Notes*; *Giving Birth to Thunder*; *Of Wolves and Men*; *River Notes*; *Crossing Open Ground*; *Crow and Weasel*; and *The Rediscovery of North America*.

Miriam Therese MacGillis, O.P., is a member of the Dominican Sisters of Caldwell, New Jersey. She is the director of Genesis Farm, a learning center where people come to search for authentic ways to live in harmony with the natural world and each other. She coordinates programs exploring the work of Thomas Berry as he has interpreted the New Cosmology. Sister Miriam has conducted over 800 workshops and seminars internationally.

Joanna Macy is a teacher and an activist in movements for peace and justice. Thousands of people have participated in her workshops in many countries. She is a co-founder of the International Interhelp Network, a global alliance of people from all walks of life who strive to integrate political, emotional, and spiritual dimensions of the work for peace. She is the author of several books including *Despair and Empowerment in the Nuclear Age* and *World as Lover, World as Self*.

Peter Matthiessen is one of the world's leading nature writers. His many books include: *The Snow Leopard* which won the National Book Award; *Baikal—Sacred Sea of Siberia*; *Cloud Forest*; *In the Spirit of Crazy Horse*; *At Play in the Fields of the Lord*; *Far Tortuga*; *African Silences*; *The Tree Where Man Was Born*; and *Nine-Headed Dragon River*.

Deena Metzger is a novelist, poet, essayist, healer, and educator concerned with the ecology of the spirit. Her recently published works include: *What Dinah Thought*; *Looking for the Faces of God*; *The Woman Who Slept with Men to Take the War out of Them*; and *Writing for Your Life: Creativity, Imagination and Healing*.

Joel Monture, Upper Mohawk, Wolf Clan, is a professor of traditional arts at the Institute of American Indian Arts in Santa Fe, New Mexico. A consultant in Native American curriculum, Monture is the author of *The Complete Guide to Traditional Native American Beadwork* and is currently completing a history of the northeast Native nations. He is also the author of a collection of short stories about Native American children.

Thomas Moore, Ph.D., is a writer and psychotherapist living with his family in New England. A leading lecturer in the area of archetypal psychology, he has specialized in the interdisciplinary study of religions, psychology, and the arts. His books include: *Rituals of the Imagination*; *The Planets Within*; the bestselling *Care of the Soul*; and *Soul Mates*.

John A. Murray is the author of twelve books including *Wild Africa*; *The Islands and The Sea*; and *A Republic of River*. He has authored over one hundred articles, essays, and reviews. He is the director of the graduate professsional writing program at the Unviersity of Alaska, Fairbanks.

Ingrid Newkirk is chairperson of People for the Ethical Treatment of Animals, the largest animal rights organization in the United States. She is author of *Save the Animals! 101 Easy Things You Can Do*; *Kids Can Save the Animals! 101 Easy Things To Do*; *The Compassionate Cook*; and *Free the Animals! The Untold Story of the U.S. Animal Liberation Front and Its Founder, "Valerie"*.

David Petersen is the author of *Among the Elk* and other books. His essays have appeared in *Northern Lights*, *Wilderness*, and three anthologies of nature writing. Petersen edited and introduced *Confessions of a Barbarian: The Journals of Edward Abbey*. His book on the ghost grizzlies of Colorado will be released in 1995.

David Rothenberg is Assistant Professor of Humanities at the New Jersey Institute of Technology. He translated Arne Naess' *Ecology, Community, and Lifestyle*, and is the author of *Is It Painful to Think? Conversations with Arne Naess*, and the editor (with Peter Reed) of *Wisdom in the Open Air: The Norwegian Roots of Deep Ecology*. His most recent book, *Hand's End*, treats of the philosophy of technology. Rothenberg is also a noted composer and jazz clarinetist, and his recording *nobody could explain it* is available on the Accurate label.

John Seed is founder and director of the Rainforest Information Centre in Australia. He has traveled to many countries lecturing and showing films to raise awareness of the plight of the rainforests, about which he has also co-produced a film. He is the co-author (with Joanna Macy, Pat Fleming, and Arne Naess) of *Thinking Like a Mountain—Towards a Council of All Beings*.

B. D. Sharma, an accomplished mathematician, is one of India's foremost thinkers on the subject of sustainable development. A Gandhian by conviction, he has been working untiringly toward the liberation of the indigenous people of India from the development trap into which they seem so inexorably to be falling. One of the most unique and unlikely weapons he uses is, ironically, the Indian Constitution, which guarantees tribal people the rights they are too often denied.

David Steindl-Rast, O.S.B., Ph.D., holds degrees from the Vienna Academy of Fine Arts and the Psychological Institute at the University of Vienna. After years of training in philosophy, theology, and the 1,500-year-old Benedictine monastic tradition, Brother David practiced Zen with many Buddhist masters. His books include: *Gratefulness, the Heart of Prayer* and *A Listening Heart*.

Terry Tempest Williams is naturalist-in-residence at the Utah Museum of Natural History in Salt Lake City. Her first book, *Pieces of White Shell: A Journey to Navajoland* received the 1984 Southwest Book Award. She is also the author of *Coyote's Canyon*; *Refuge: An Unnatural History of Family and Place*; and two children's books.

—About the Editors—

Michael Tobias is the author of twenty books, and the writer, director, and producer of more than seventy films. He is best known for his ten-hour dramatic miniseries and novel, *Voice of the Planet*, and his most recent books, *A Naked Man*, and *World War III—Population and the Biosphere at the End of the Millenium*.

Georgianne Cowan is the director of the Spirit and Nature program at the Earth Trust Foundation. For fifteen years she has been teaching workshops on the creative feminine, women and nature, and classes in subtle and expressive movement. Her photography has been widely published, and the internationally distributed video, *Earth Dreaming*, is based on her photo imagery.

DUTTON **PLUME**

AMERICA'S READING ABOUT—

☐ **OUR STOLEN FUTURE** *Are We Threatening Our Own Fertility, Intelligence, and Survival?—A Scientific Detective Story* **by Theo Colborn, Dianne Dumanoski, and John Peterson Meyers.** Man-made chemicals have spread across the planet, permeating virtually every living creature and the most distant wilderness. Only now, however, are we recognizing the full consequences of this insidious invasion, which is derailing sexual development and reproduction, not only in a host of animal populations, but in humans as well. The authors explore what we can and must do to combat this pervasive threat. (939822—$24.95)

☐ **EARTH IN THE BALANCE** *Ecology and the Human Spirit* **by Vice President Al Gore.** An urgent call to action to save our seriously threatened climate, our water, our soil, our diversity of plant and animal life, indeed our entire living space. "A powerful summons for the politics of life and hope."—Bill Moyers
(269350—$13.00)

☐ **NAKED EARTH** *The New Geophysics* **by Shawna Vogel.** In a beautifully crafted work that is at once impeccably authoritative and wonderfully accessible, the author provides readers with new explanations to ancient mysteries and the latest geophysical hypotheses. This book offers an illuminating and thoroughly riveting account of the pioneering geophysicists of today. (937714—$10.95)

☐ **THE SOUL OF NATURE** *Visions of a Living Earth.* **Edited by Michael Tobias and Georgianne Cowan.** Gathered in this diverse anthology are some of the most cherished and eloquent voices from the eco-spirituality movement. It includes more than 30 essays—both passionate and reflective—from pivotal writers and thinkers of our time such as, Annie Dillard, David Peterson, Thomas Moore, Barry Lopez, Gretel Ehrlich, and others. It is a plea on behalf of the biosphere and a blue-print for an ecological spirituality for the new millennium. (275733—$11.95)

Prices slightly higher in Canada.

Visa and Mastercard holders can order Plume, Meridian, and Dutton books by calling
1-800-253-6476.
They are also available at your local bookstore. Allow 4-6 weeks for delivery.
This offer is subject to change without notice.

℗ **PLUME**

WORLD ISSUES

☐ **DON'T BELIEVE THE HYPE** *Fighting Cultural Misinformation about African-Americans* **by Farai Chideya.** This book is filled with factual ammunition for fighting the stereotypes and misinformation too often accepted as the "truth" about the 31 million African-Americans in this country. The author draws on hard fact, not hype to show the real picture on key subjects like jobs, education, social welfare, crime, politics, and affirmative action. (270960—$10.95)

☐ **WHICH SIDE ARE YOU ON?** *Trying to Be for Labor When It's Flat on Its Back* **by Thomas Geoghegan.** This gripping epic of economic loss and shattered dreams demands we take a fresh look at the dilemmas of class in modern America, and at ourselves. "Brilliant, inspiring . . . charming and acidic at once . . . unparalleled in the literature of American labor."—*The New York Times Book Review* (268915—$11.95)

☐ **THE LITIGATION EXPLOSION** *What Happened When America Unleashed the Lawsuit* **by Walter K. Olson.** From malpractice suits to libel actions, from job discrimination to divorce, suing first and asking questions later has become a way of life in the United States. Here is the first major exploration of this trend—why it developed, who profits and who loses, and how it can be contained. (268249—$13.95)

☐ **THE AGE OF MISSING INFORMATION by Bill McKibben.** In this brilliant, provocative exploration of ecology and the media, McKibben demolishes our complacent notion that we are "better informed" than any previous generation. "By turns humorous, wise, and troubling . . . a penetrating critique of technological society."—*Cleveland Plain Dealer* (269806—$10.95)

Prices slightly higher in Canada.

Visa and Mastercard holders can order Plume, Meridian, and Dutton books by calling
1-800-253-6476.
They are also available at your local bookstore. Allow 4-6 weeks for delivery.
This offer is subject to change without notice.

Ⓟ **PLUME** **DUTTON**

TIMELY ISSUES

☐ **A GARDEN OF UNEARTHLY DELIGHTS** *Bioengineering and the Future of Food* **by Robin Mather.** In this first definitive book in the controversial subject of bioengineering and food, the author explores the implications of the new cost-efficient, flavor-enhanced creations about to move out of the laboratory and into your supermarket. This balanced view takes in the whole spectrum of ideas on this charged topic, from genetic alteration at one end to small-farm sustainable agriculture at the other. (272637—$11.95)

☐ **BEYOND BEEF** *The Rise and Fall of Cattle Culture* **by Jeremy Rifkin.** This persuasive and passionate book illuminates the international intrigue, political give-aways, and sheer avarice that transformed the great American frontier into a huge cattle breeding ground. "Should be compared with *Silent Spring, The Fate of the Earth,* or *Diet for a Small Planet*. . . . Draws our attention to a threat to what we most value."—*New York Review of Books* (269520—$12.95)

☐ **THE MAN WHO GREW TWO BREASTS** *And Other Tales of Medical Detection.* **Berton Roueché.** Here in this book are some of the most intriguing and enlightening true tales of medical mystery and detection researched and re-created by the foremost medical journalists of our time. You will come to understand the human factor in the scientific mosaic that is medicine, while enjoying the excitement of following a mind-teasing trail of clues in pursuit of a tantalizingly elusive quarry. (939342—$22.95)

Prices slightly higher in Canada.

Visa and Mastercard holders can order Plume, Meridian, and Dutton books by calling
1-800-253-6476.
They are also available at your local bookstore. Allow 4-6 weeks for delivery.
This offer is subject to change without notice.

Ⓟ **PLUME** **MERIDIAN**

CRITICAL THINKING

☐ **BEYOND CRISIS** *Confronting Health Care in the United States.* **Edited by Nancy F. McKenzie. Foreword by Barbara Ehrenreich.** This timely work provides an unflinching and comprehensive survey of the current state of American health care delivery. There are full examinations of the array of proposals for reform now on the table, including an in-depth look at the Clinton administration's proposed reforms, explorations of the single-payer alternative, as well as selections devoted to community activism and innovations in health care delivery.
(011086—$19.95)

☐ **AS REAL AS IT GETS** *The Life of a Hospital at the Center of the AIDS Epidemic.* **by Carol Pogash. Foreword by Randy Shilts.** "An exciting cram course about an invidious disease and about politics and human behavior. The human dimensions of the AIDS epidemic, 'the most important medical story of the century,' grip the reader who comes to know the hospital's doctors, nurses, and patients in this remarkable book."—*Library Journal*   (271274—$9.95)

☐ **TAKING CARE OF OUR OWN** *A Year in the Life of a Small Hospital.* **by Susan Garrett.** A wonderful blend of heartwarming stories, high drama, and sensible, well-informed discussion on the American health care crisis, this is a mind-changing, idea-generating book that offers workable solutions and a compassionate perspective that caring citizens and enlightened lawmakers need, now more than ever, to hear. "Insightful . . . a small gem of a book." —*Washington Post Book World*   (272718—$10.95)

Prices slightly higher in Canada.

Visa and Mastercard holders can order Plume, Meridian, and Dutton books by calling
1-800-253-6476.
They are also available at your local bookstore. Allow 4-6 weeks for delivery.
This offer is subject to change without notice.